RACE, REASON
and
RUBBISH

*A Primer of Race Biology
for the Plain Man*

By the same author

STATISTICAL METHODS FOR
MEDICAL AND BIOLOGICAL STUDENTS

An Examination of the Biological
Credentials of the Nazi Creed

Race, Reason
and
Rubbish

by
GUNNAR DAHLBERG
M.D., LL.D.

*Director of the State Institute of Human
Genetics in the University of Uppsala*

Translated from the Swedish by
LANCELOT HOGBEN, F.R.S.

*Regius Professor of Natural History in the
University of Aberdeen*

NEW YORK
COLUMBIA UNIVERSITY PRESS
1943

FIRST PUBLISHED IN 1942

ALL RIGHTS RESERVED

PRINTED IN GREAT BRITAIN
in 12-Point Fournier Type
BY BILLING AND SONS LIMITED
GUILDFORD AND ESHER

CONTENTS

CHAPTER		PAGE
	Translator's Foreword	11
	Author's Preface	15
1.	Fundamental Mechanisms of Inheritance	19
2.	Independent Assortment and Linkage of Genes	47
3.	Polymeric Characters, Dihybrid Inheritance and Mutation	74
4.	Sex Determination and Sex-linkage	94
5.	Environment and Gene Exhibition	117
6.	Principle of Random Mating	132
7.	Selection	148
8.	Inbreeding and Cousin Marriage	163
9.	Assortative Mating	173
10.	The Isolate Effect	180
11.	Race	193
12.	The Jewish Question—Conclusion	219

LIST OF ILLUSTRATIONS

FIG.		PAGE
1.	Fertilisation of the egg of a marine organism in which the egg itself is relatively small and divides as a whole	19
2.	Diagram showing Darwin's conception of inheritance (left) and the modern view (right)	22
3.	Diagram of cell lineage showing the separation of the germ track at the beginning of development	23
4.	Diagram of cell lineage in the human embryo	25
5.	Inheritance of pigment in a cross between negro–white half-castes as conceived in terms of the old Dilution theory	28
6.	Inheritance of pigment in a cross between negro–white half-castes as we now understand the facts	29
7.	Cross of Ivory and Red Snapdragons illustrating Mendel's Laws	31
8.	Crossing of black and white Andalusian fowls to illustrate Mendel's Laws	35
9.	Simple mendelian inheritance of the *recessive* white and *dominant* black coat colour in mice	37
10.	Photomicrographs of the first division of the egg of the Horse Thread Worm showing the four chromosomes	41
11.	Diagrammatic representation of normal nuclear division in the growing tissues (including young germ-cells) above, reduction division before formation of sperm below	42
12.	To illustrate that a recessive gene (V) may remain concealed for many generations of mating till two heterozygous individuals (SV) chance to mate	48

Race, Reason and Rubbish

FIG.		PAGE
13.	Transmission of two recessive characters of the Banana fly—mating of the recessive *ebony* mutant with the recessive mutant having *vestigial* wings	51
14.	Chessboard diagram to illustrate recombinations of two independent pairs of genes, i.e. genes located on different chromosomes	53
15.	Cross between two races of Antirrhinum (snapdragons) distinguished by several genes affecting shape and colour of the flower	56
16.	Multiple effects due to a single gene	67
17.	Sex determination in the Banana fly. Chromosomes in normal cell division of male (right) and female (left) top row, and gametes in middle row	69
18.	Diagram to show crossing-over of genes owing to twisting of chromosome when the pairs separate at the reduction division	70
19.	Photomicrograph of part of a chromosome of the salivary gland of the Banana fly, showing dark stripes to each of which identified genes can now be assigned	72
20.	Part of one of the chromosomes of the salivary gland of the Banana fly—the dark stripes correspond to the situation of genes which are known	73
21.	Inheritance of Comb shape in Poultry	78
22.	Diagram of the inheritance of the Lethal yellow coat colour of mice	81
23.	Chromosome Mutations of Tomatoes. The three varieties are distinguished by possessing 24, 36 and 48 chromosomes	84
24.	The Evening Primrose (*Oenothera Lamarkiana*)	85
25.	A chromosome mutation in the Banana fly. A segment of one member of one pair is attached to a member of another pair equally mated in cells of the normal fly as seen on the right	88

Race, Reason and Rubbish

26. The normal hybrid (left below) *Primula kewensis* produced by cross polination of *Primula foribunda* (left above) and *Primula verticellata* (right above) is sterile; but a chromosome mutant of *Primula kewensis* with double the normal number of chromosomes is a new fertile species 89

27. Vegetative Reproduction of the Fresh Water Polyp *Hydra*. The organism consists of a contractile trunk attached at one end to a solid object. The mouth is at the free end surrounded by grasping tentacles 92

28. Chromosomes from the cells of the male (above) and female (below) of a plant bug (*Protenor*) showing the unpaired X-chromosome of the male 96

29. Sex determination in birds—the female is the heterogametic sex 98

30. Mosaic Wasp gynandromorphs. The dark regions have male characteristics, the light ones have the peculiarities of the corresponding structures in a female 100

31. Castrated pullet with male spurs and sickle (above) and capon (castrated cockerel) with reduced comb but otherwise typically male 102

32. Common foetal circulation of cattle twins 104

33. Crossing-over in transmission of two recessive sex-linked characters of the Banana fly—mating of pure wild type female with red eyes and gray body to double recessive male with white eyes and yellow body 105

34. Crossing-over in transmission of two recessive sex-linked characters of the Banana fly; mating of yellow male with wild type (red) eyes to white-eyed female with wild type (gray) body colour 107

35. Sex intergrades of the Chocolate moth showing various graduations between the pale female (*a*) with thick abdomen and filamentous antennae to the dark male (*k*) with thin abdomen and feathered antennae 111

Race, Reason and Rubbish

FIG.		PAGE
36.	Diagram to represent Goldschmidt's interpretation of sex differentiation	113
37.	Diagrams showing how unequal distribution of extra-nuclear cell contents leads to the production of cells in which *identical* chromosomes react with a different substratum of other material	122
38.	Normal Russian rabbit (above) with black pigment confined to the extremities, and individual of the same variety kept in cold after shaving the regions where the new hair is now black	124
39.	Diagram to illustrate Johannsen's experiments on beans (p. 125) showing that selection within pure lines has no effect on the character of the offspring	126
40.	Uniovular (monozygotic) twins	128
41.	Diovular (dizygotic) twins	129
42.	Graph to show how the frequency of a character varies with the frequency of the genes	141
43.	The Effect of Selection. The heights correspond to frequency of recessives in successive generations, when all recessive individuals in each generation are prevented from having any offspring	150
44.	Graph showing percentages of each year-group for the school leaving examination (p. 154). Males (above) and females (below) are plotted separately	155

Translator's Foreword

THIS book deals with biological aspects of the race problem, and the Scandinavian antecedents of the author give a peculiar piquancy to what he has to say about the suppositious superiority of the Nordic Man. Professor Dahlberg is one of the six living people who know most about heredity. He is unique in bringing to his work the severe discipline of Swedish medical education, a sound grasp of statistical method, humane judgement and horse sense. As head of the State Institute of Race Biology and, at the time of writing, Dean of the Medical Faculty in the University of Uppsala, he is recognised as the leading authority on Human Genetics and Medical Statistics in the Scandinavian countries. Of the esteem which his work enjoys elsewhere, it is enough to say that the University of Aberdeen has conferred on him an honorary LL.D. degree. Needless to say, Aberdeen is not notorious for what it gives away for nothing.

Perhaps the circumstances in which the book was translated into the Anglo-American language are almost as unusual as the gifts of the author. Two days before Hitler's unexpected aggression in Norway, when I was still staying with Gunnar Dahlberg and his talented wife in Uppsala, we had discussed the possibilities of translating this book into the international language of the United States and the British Commonwealth of

Race, Reason and Rubbish

Nations. The night before the Nazis entered Oslo, I lectured on the methods of human genetics in the Anatomical Institute of the University of Oslo, and had intended to fly to Copenhagen next day. *En route* to the aerodrome with my daughter Sylvia, who typed the first draft of this translation, the bombs began to fall. We saw the Nazi army march down Karl Johan Gate in the afternoon. Then we took the advice of the American Consul and legged it out of town without our baggage. The British Legation had already set us a good example.

When we arrived in a lorry at the Swedish frontier, Gunnar Dahlberg's name put us right with officialdom. Thereafter I had to kick my heels (in his socks), cut from any connexion with Britain or America. Since I am not streamlined for indolence, it was gratifying to fill in the first fortnight by converting this book from a language spoken by less than seven, and understood by less than fourteen million, into one which is spoken and understood by about four hundred and ninety million human beings.

My modest task as a translator has been simplified by the fact that my own scientific judgement coincides closely with the author's conclusions, especially with reference to the main thesis which deals with what he has called *isolates* in human populations. Inevitably scientific workers disagree about matters which lie at the interface between proof and discovery; and it would be a dull world were it otherwise. I hope that where my views do not coincide with the author's I have interpreted his intentions conscientiously. Fortunately for both of us, Professor Dahlberg's command of English is much greater than my limited grasp of Scandinavian languages.

LANCELOT HOGBEN

UPPSALA
April 1940

Postscript.—The cordial thanks of the author and translator are due to Dr. J. Sang, who has seen this book through the press.

Author's Preface

ONE cannot discuss race problems without proper knowledge of the laws of heredity. Any account of modern genetics in its application to mankind must also leave the reader dissatisfied if he or she does not get to know something about the race problem. So this book is partly made up of chapters dealing with general aspects of heredity, followed by a series of others about the nature of human inheritance. These lead up to the discussion of race.

It is always more or less uncongenial and somewhat ridiculous to a man of science when matters bearing on his own field are transferred to the plane of party politics, and become the topic of excited controversy among a wider public. To some extent discussion must then descend to a lower level. In any collection of people some individuals are apt to arrive at very decided conclusions from an extremely narrow basis of information. Anyone of this type, having formed an opinion, knows how things stand and has no inclination to procure further enlightenment on his own behalf, though he often has a strong urge to make propaganda for his beliefs. In many circumstances public discussion about problems of a scientific nature therefore evokes a sentiment of repugnance in a man of science.

It must have been extremely awkward for astronomers

Race, Reason and Rubbish

and physicists when laymen eagerly debated the theories of Copernicus and when his opinions were condemned by the Roman Catholic Church in the sixteenth century. The frantic discussion which followed the announcement of Darwin's doctrine now seems quite remarkable, and when a legal action is taken against a schoolmaster in a small American town because he has given instruction about evolution it raises a solitary smile. Time will come when public discussions, such as those we now have about race, will be looked upon as equally queer.

Meantime the problems of human inheritance are important to posterity; and it is natural that we should be interested in them. Even if we do consider public discussion of particular scientific issues regrettable from some points of view, it is unavoidable and for other reasons useful and necessary to discuss them. Scientific men have no claim to rank as a caste with the peculiar privilege of pronouncing judgement on difficult questions. The public has a right to be aware of, as well as to participate in, the consequences of scientific knowledge. What alone is important is that argument should be free, and that even men of science may themselves have the opportunity of stating their opinions without restriction. To prevent discussion from degenerating, it is desirable for the public to be able to get hold of elementary information. It is only such that I wish to transmit in the compact survey contained in this book. As far as zoology and botany are concerned, there is

Race, Reason and Rubbish

already a good literature of popular scientific exposition concerning research about inheritance. Partly because those who write books about inheritance have been zoologists and botanists, this is not true of human genetics. At the same time human beings differ in so many ways from other species that a succinct popular exposition should be particularly appropriate to-day.

GUNNAR DAHLBERG

UPPSALA

August 1939

CHAPTER I

In ordinary circumstances an egg develops after it has first been fertilised, that is to say, after a sperm (Fig. 1) has made its way into it. At the time it produced a very big sensation when it was first shown that one could get unfertilised eggs to develop into full-grown individuals. This result was brought about by treating eggs of some animals with chemical reagents. One can even

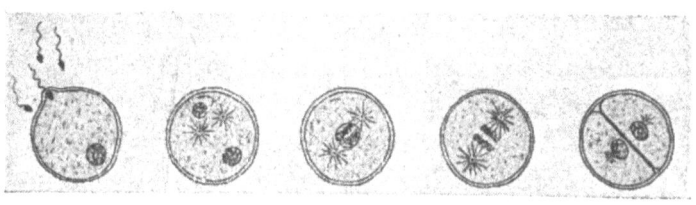

Fig. 1.—Diagram of the fertilisation of the egg of a marine organism in which the egg itself is relatively small and divides as a whole

give an egg the impetus to development by wounding it with a delicate needle. So it might be suspected that the sperm brings about the development of the egg by the minute prick produced at its entry. Later researches have shown that pricking with needles which are small enough to answer for the sperm itself have no effect at all. The result can only be got with thicker ones. By such means it is possible to bring about the development of the egg of such highly organised animals as frogs,

Race, Reason and Rubbish

and the full-grown individual does not visibly differ from an adult which develops from a fertilised egg. It has also been possible to carry out similar experiments with sperm. One must then supply a portion of cell substance from an egg, because the sperm has insufficient food reserves for further development.

From such experiments we can draw the conclusion that both egg and sperm have a full equipment of hereditary materials needed for building up an individual. In ordinary circumstances, i.e. when an individual develops after fertilisation, the adult must therefore have a duplicate set. These pairs of hereditary *determinants* may consist of components which are quite alike, but may also consist of members which are more or less different. In genetic research one tries to find out the effect of dissimilarities among the hereditary potentialities or *determinants* and to decide what happens when they are combined in different ways.

Before we go into a discussion of such questions, we ought to try to get a grasp of what is meant by a hereditary determinant. In the language of everyday life we use such words as *hereditary* or *inherited disposition* in a very loose way. As a rule this does not give rise to any inconvenience, but if we want scientific insight into the problem of inheritance we have to get at a clearer definition of what we mean by such expressions as *hereditable* properties and the like. Strictly speaking, when we say someone inherits brown eyes from father or mother, the

Race, Reason and Rubbish

expression is incorrect. We do not inherit eye colour. We never inherit the character itself but only the potentiality for producing it. This implies the possibility that we inherit the potentiality for a character which may not necessarily be found in either of our parents. Potentialities can exist without conferring on a particular individual any corresponding characteristic.

In reality it took a long time before people were clear about this. Even in Darwin's time, no distinction was made between character and disposition. Darwin supposed that in some way we inherit characters directly. He thought that small particles wander from different parts of the body to the sex cells (Fig. 2). Migrating particles from the eye, for example, might carry to them—among other things—eye colour. At the time the theory furnished an explanation of how an effect upon an organ of one of the parents might be able to bring about a change of hereditary disposition in such a way that acquired characters could be transmitted. In Darwin's time people believed that this was really so. His theory is less attractive when we take into account the fact that we should have to postulate an unprecedented quantity of such particles on the move.

Since we have begun to analyse more deeply how fertilised eggs develop, the so-called inheritance of acquired characters has become even less plausible. When a fertilised egg develops, it divides into two cells. Among other things one of these cells will even-

Race, Reason and Rubbish

tually produce sex organs and germ-cells (gametes). The other is going to produce quite different organs.

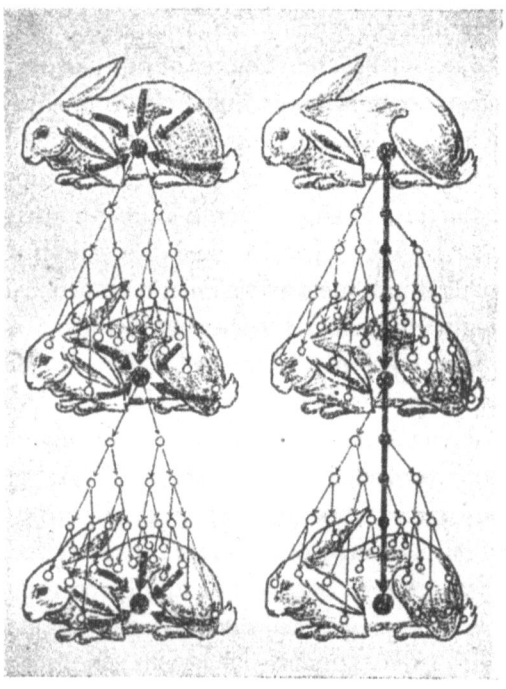

FIG. 2.—Diagram showing Darwin's conception of inheritance (left) and the modern view (right) that the body of an individual is an offshoot of the germ-track which continues from one generation to another

When each of these cells subsequently divide, the first will again split into two. Again, only one of them is destined to produce gonads. In this way there is a distribution of cells (Fig. 3) which are all descendants of the egg, so that some of them become germ-cells and others

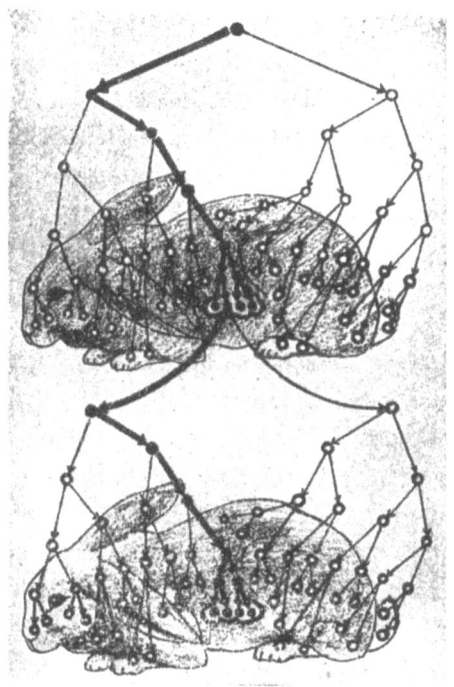

FIG. 3.—Diagram of cell lineage showing the separation of the germ-track at the beginning of development

contribute to other sorts of tissues. The germ cells are descended directly from the egg, and all other cells which make up the other organs of the body are relatives belonging to another branch of the same family tree. In the horse thread worm (*Ascaris megalocephala*) there is a visible difference between the cell-lineage which ends in the formation of the gametes, i.e. the so-called *germ-track*, and the cell-lineage which ends in the formation of the rest of the body, i.e. the so-called *somatic-track*. The germ-track has cells with nuclei which look different from those of the somatic-track. With the microscope it is possible to see directly how the clearer ones gradually become body cells and the darker become cells of the sex organs.

Among higher animals one can even distinguish between three branches (Fig. 4) of this family tree of cell descendants. When the fertilised egg of a human being develops, a group of cells become embryonic membranes such as the placenta. It has been called the *trophoblast-track*. This region of the fertilised egg forms a protective organ which is essential to the development of the egg during embryonic life. When it has done its work the whole cell group is thrown off and dies. Another portion of the egg builds the *soma*, that is to say, all the organs which collectively make up an individual with the exception of the sex cells. This somatic-track encloses and protects the germ-track, and has the job of securing food, providing shelter against injury and preparing for the possibility of reproduction. Sooner or later it will die. The germ-track itself produces a succession of cells of which the majority will also die;

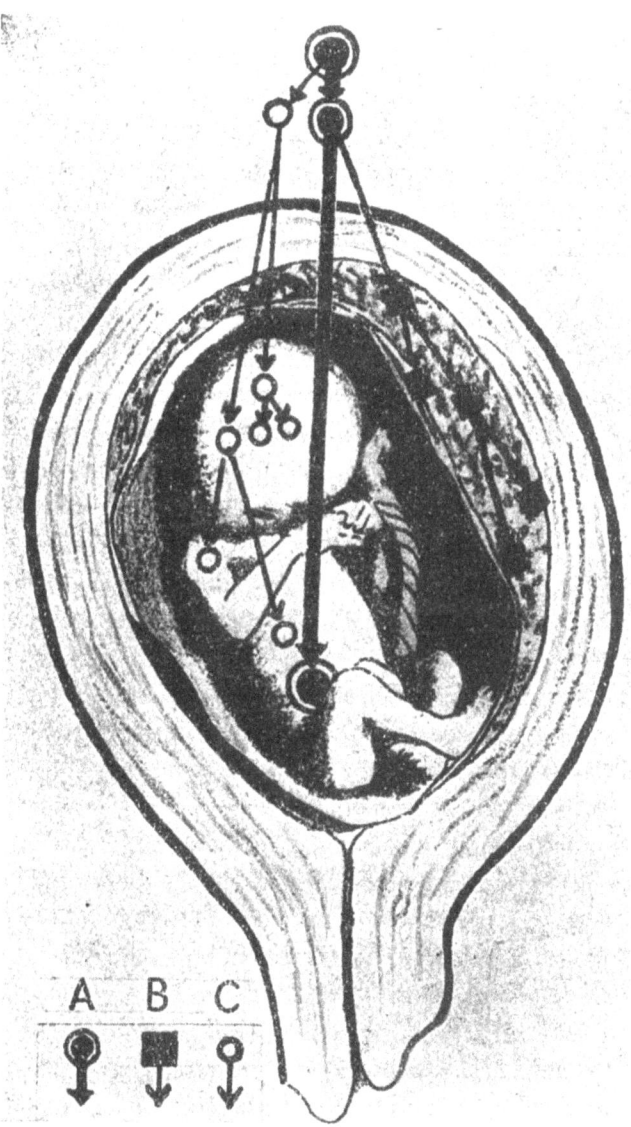

FIG. 4.—Diagram of cell lineage in the human embryo:
(a) the *germ*-track, (b) the *trophoblast*-track, (c) the *somatic*-track.

but some of them may attain union with other sex cells. In this way they pass over into a new being. So the germ-track is the only path to life everlasting.

From this point of view it is commonplace that hereditary potentialities pass on from individual to individual and that their contribution to characteristics which collectively make up what we call individuals is an offshoot of the germ-track itself. Meanwhile the germ-track goes on living, but only through perpetual reunion with other germ-tracks. Potentialities of one individual get mixed with those of others, so that new individuals which turn up may get a very diverse equipment of hereditary materials. With respect to this equipment every individual is a Jekyll and Hyde composed of potentialities derived from both parents. It goes without saying that a new individual is sometimes more like one of its parents, sometimes more like the other, and not identical with either.

At first sight this statement of the case faces certain difficulties. Suppose, for instance, that there are 1,000 hereditary determinants in every cell of a single individual. If all of them are present in each of its germ cells, and if every new individual results from the fusion of two cells, the fertilised egg should get 2,000. When the individual which develops from it in turn produces eggs which are also fertilised, the cells of the offspring would have 4,000. Thus a mechanism of this sort *apparently* implies that the hereditary determinants get doubled in each generation. Even if every one of them is a tiny particle, this still means that the total mass of hereditary materials is doubled. In the long run there would be no more available space in the small sex cells.

Hence the fact that the germ cells do not increase in size in each generation must mean that half of the hereditary determinants play dummy in the formation of new ones. Ripe eggs and sperm must contain a single complement, while the individual which develops from a fertilised egg must have a double stock of hereditary material. So we call the equipment of the ripe germ cells *haploid*, i.e. single, and that of the fertilised egg *diploid*, which means double. Strictly speaking, we might say that there is a sort of alternation of generations among higher organisms. Adults with a duplicate stock of hereditary material live a comparatively long time and make way for a shortlived generation of germ cells with a single dosage.

Considerations of this sort led up to what is called *Mendelism*, i.e. the doctrine which was put forward during the 'sixties of the last century by the Austrian monk, Gregor Mendel. He announced his discoveries in a little-known journal. Partly for this reason, and also perhaps because he was not a professional man of science, his discoveries were neglected at the time. They conflicted with the then prevalent outlook according to which the material basis of inheritance is a homogeneous substance which can be diluted and mixed as we can mix

Race, Reason and Rubbish

whisky and soda. Accordingly there would be a blending (Fig. 5) of both sorts of hereditary material when a

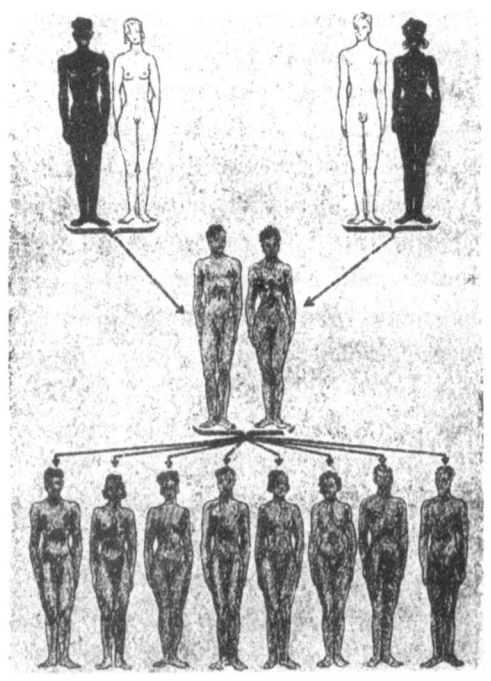

FIG. 5.—Inheritance of pigment in a cross between negro–white half-castes as conceived in terms of the old Dilution Theory which people believed in before Mendel's work was recognised

negro or a negress mate with a white woman or a white man. The result is, that the issue would be lighter than the former and darker than the latter. Naturally, there-

Race, Reason and Rubbish

fore, it would be a mulatto. If the latter mated with a white there would be further dilution with the production of an individual lighter than the mulatto parent. To some

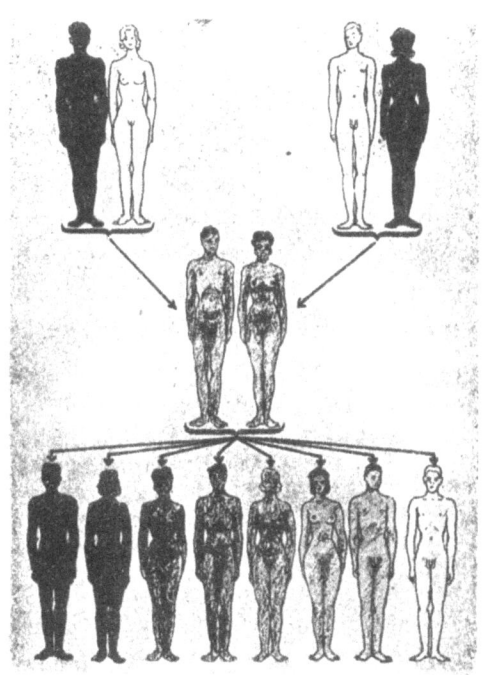

Fig. 6.—Inheritance of pigment in a cross between negro–white half-castes as we now understand the facts

extent the experiment had been made and sustained expectation in agreement with this point of view, but anomalies had also been found. Crossing sometimes

Race, Reason and Rubbish

(Fig. 6) produces individuals who have skin as black as that of a negro. People then spoke of a reversion to the hereditary constitution of an older generation, but the fact that they could christen the phenomenon did not imply real understanding.

One reason why is was difficult to elucidate the problems of inheritance was that people tried to disentangle complicated crosses and conscientiously investigated all the differences between parents and offspring. Mendel himself concentrated his enquiries on simple characteristics. He used chiefly peas for his experiments and tried to disentangle the results when he crossed strains which differ among themselves in one respect only. By limiting the problem in this way to two characteristics he was able to make experiments which were simple to grasp. Mendel carried out his researches in the monastery garden in Brunn. He was a mathematical teacher in the monastery school and had a robust intelligence with a mathematical bias. In spite of the scanty recognition he earned, he followed out his experiments far enough to lay a firm foundation for genetic research. The results Mendel obtained, and the theories he put forward, are permanently fundamental for the science of heredity; but his work did not begin to exert influence until 1900. In that year the Mendelian laws were rediscovered by three investigators: Correns, de Vries, and Tschermak.

Mendel assumed that the hereditary determinants are

Race, Reason and Rubbish

indivisible entities, that the material basis of heredity is built up of such particles, and that therefore it is not a substance which can be diluted. This involves an intellectual step forward with as much significance for

From HEREDITY *by A. F. Shull*

FIG. 7.—Cross of Ivory and Red Snapdragons illustrating Mendel's Laws

research into inheritance as the atomic theory had for chemistry. Just as atoms are material entities which cannot be split without changing their nature and can never be destroyed, hereditary determinants are entities which, according to Mendel, can be present as a whole or entirely absent, but cannot be present in part. If the material basis of heredity really consists of such entities, and if we also assume that these entities combine at random in much the same way as when we toss for heads or tails, we can calculate what results follow crossing.

To demonstrate Mendel's laws we may turn our attention to crosses between snapdragons (Fig. 7). If we cross a white snapdragon with a red snapdragon we get pink ones. The white snapdragon has a double stock of determinants and hence two determinants responsible for the white colour. The red one has two for the red colour. When the germ-cells are formed we have only to reckon with one representative of each pair. Thus the red snapdragon forms germ-cells with one determinant for red colour. The white snapdragon forms germ-cells with one for white. When germ-cells of both sorts unite, they produce an individual which has a determinant for red colour and one for white colour. The two of them may be said to make a compromise, so that the new plant is midway between its parents.

Up to this point Mendel's theory is in agreement with the older theory of dilution. If the latter is true, we should expect to go on getting pink ones if we cross the pink snapdragons with one another, but according to Mendel's hypothesis the result would be otherwise. Each pair of such determinants in the cells of a plant with pink flowers consists of one for red and one for white,

Race, Reason and Rubbish

and we have only to reckon with one of them. The pink snapdragons form some germ-cells which have the determinant for white. If chance decides which determinant is going to be in a particular germ-cell we ought to get germ-cells with the determinant for red or for white equally often, as we get a head or a tail with equal frequency when we toss a coin. So if the germ-cells unite at random we have three possibilities. Two germ-cells with the determinant for red colour can meet one another. We then get an individual which has a double equipment of the determinant for red pigment and therefore has red flowers. Two germ-cells with the determinant for white may fuse. The individual so produced must then have white flowers. Finally, a germ-cell with the determinant for red colour may unite with one which has the determinant for white. The individual which then develops ought to produce pink flowers.

When we actually carry out crosses we get these three different types of individuals, and we find that one quarter of the plants have white flowers, one half pink, and one quarter red. At first sight it seems odd that the pink ones should be as numerous as the red and white ones put together, but if there is haphazard fusion among the germ-cells, this proportion is just what we should expect. If we toss for heads or tails twice in succession we have the possibility of getting heads in the first as well as the second toss, the possibility of getting tails in both, the possibility of getting a head at the first followed

Race, Reason and Rubbish

by a tail, and the possibility of getting first a tail and then a head. In other words, the probability of getting both a head and a tail is twice as great as getting two heads in succession or two tails in succession. This is so, because we can get the same result in two ways.

So if we use simple calculations based on probability we should expect to get exactly the result which we do get when we cross pink snapdragons. We ought to get offspring of which one quarter are snapdragons with red, one quarter with white, and one half with pink flowers. This proportion, and the principle which lies behind it, is called Mendel's Law. We can get a further check on the latter by continuing the crosses. If we cross the white offspring with one another we ought to get only white snapdragons. In everyday language the white ones are what we call *thoroughbred* or *true to type*. The same is true of the red snapdragons. But if we cross the pink ones with one another we ought to go on getting a splitting up into one quarter white, one quarter red, and one half pink. This is what really happens. Pink snapdragons are not true to type and can never be so.

A conclusion such as this may have practical value. For instance, if we mate white and black chickens of the Andalusian breed (Fig. 8) we get gray ones which correspond to our pink snapdragons and will thus have the determinant for black as well as for white feathering. These gray fowls are called *Blue Andalusians* and fanciers have engaged in earnest efforts to get a constant race of

Race, Reason and Rubbish

them by repeated crossing. The undertaking is doomed to failure from the start, as indeed it has failed; and with the help of Mendel's Law we could anticipate this. There is no possibility of getting a type with a characteristic which depends on two dissimilar hereditary determinants to breed true from generation to generation. If we want

From PRINCIPLES OF GENETICS *by E. W. Sinnott and L. C. Dunn.*

FIG. 8.—Crossing of black and white Andalusian fowls to illustrate Mendel's Laws

to make sure about getting blue Andalusians, the thing to do is to cross black ones with white. Then all the offspring will certainly be gray. There is no possibility of getting a fixed race of blue Andalusians breeding true to type. In crosses of plants and animals it is essential to investigate in what proportions different types turn up.

Race, Reason and Rubbish

From the figures we get we can draw conclusions about the possibilities of getting characteristics which keep constant from one generation to another. Manifestly Mendel's Law has therefore a very great significance for the practical task of improving strains, whether of plants or of animals.

To short-circuit repeated reference to individuals with dissimilar determinants and individuals with the same ones we can use special terms. Individuals with pairs of similar members are called *homozygotes*, and individuals with pairs of dissimilar components *heterozygotes*. It is true that crossing purebred varieties always produces heterozygotes which answer to a condition midway between both types of homozygotes from which they are descended. The determinants do not always arrive at such a compromise. It may also happen that one has a stronger influence than the other, so that the latter is quite overridden.

For instance, if we cross white and black mice (Fig. 9), the offspring of the first generation are black. These individuals have got a determinant for black pigment from one of their parents and for white hair from the other one; but the determinant for black pigment suppresses the one for white. The one for white is, so to say, shy, and the one for black pigment is self-assertive. In technical language we say that the black one is *dominant* and the white *recessive*. If we cross these heterozygotes which are black, but which still have a

Race, Reason and Rubbish

determinant for white as well as for black hair, one result is that we may get an individual which has two

Fig. 9.—Simple Mendelian inheritance of the *recessive* white and *dominant* black coat colour in mice (p. 36)

determinants for black pigment in each of its cells and, being homozygous for black, it always produces black offspring when mated with its like. We may also get an individual which has a double set of white determinants and thus breeds true to type. For every one of these two thoroughbred types we get two individuals which have both the determinants for black and for white. Since the black one is dominant such individuals are black. Of the total, three-quarters of their offspring will be black and one-quarter white. Among the black ones only one-third are homozygous for black. Two-thirds of the black ones are heterozygotes which, if mated with one another, produce offspring of which three-quarters are black and one-quarter white.

In order to have a convenient way of representing them we shall sometimes use letters of the alphabet as symbols for the determinants, most usually the initial letter of the name for the dominant one. Thus the determinant for black hair is represented by B. The corresponding recessive determinant is represented by the corresponding small letter. So the determinant for white hair in Fig. 9 is b. The black individuals which are homozygotes are represented by BB, black ones which are heterozygous by Bb, and white ones by bb.

When we are concerned with heterozygotes, we have to take stock of three possibilities. The determinants can come to a compromise, and we then have *intermediate* inheritance. One determinant can overpower the other, and we than have *dominance*. In addition, the heterozygote may exhibit an appearance completely different from that of both its parents. When this is so, we might speak of what happens as *extramediate* inheritance. Clearly it is a matter of taste whether we say that intermediate or extramediate inheritance takes place. When we mate black with white Andalusians and get birds which are grayish, we may regard what happens as intermediate inheritance, because we can look upon gray as a half-way house between black and white. At the same time, we can take it that the gray colour is something essentially new, the more so because it is not pure but has a bluish tint. If the colour were a bit more blue we should be fully entitled to say that it does not lie on the direct line of variation which we should expect to find between the characteristics of the parents but to one side of it. It is, in fact, a thing apart.

The three processes which we have thus described are fundamental mechanisms of inheritance, but there are special and more complicated sorts of transmission which we shall deal with in what follows. None the less, the process which lies behind the transmission of characters depends on the principles we have now examined. The Mendelian proportions and laws have been built upon logical reasoning well-grounded in experiment, and have now been checked by countless researches. Two conclusions we have been able to draw so far are as follows. Every individual has a double equipment of determinants, but only half of them pass over into the germ-cells, so that every individual gets just as many from its father or from its mother. We should now try to examine phenomena which confirm these assumptions by direct observations upon the germ-cells themselves.

Since we inherit just as many determinants from father or mother, they should be localised in such parts of the sperm and egg as are roughly equivalent. A sperm consists of a head which contains a cell nucleus of which the essential part is a deeply-staining substance *chromatin*. Besides this it has a tail which propels it forward by writhing movements. The egg also contains a nucleus which itself contains chromatin and lies in a large cell body composed of albuminous protoplasm with a larger or smaller quantity of food material. Since the *cell nucleus* is apparently that part of the cell common to egg and sperm, it is plausible to expect that the determinants are in it. It is not so reasonable to assume that they are in the *cell body*, which is characteristically bigger in the egg than in the sperm. So we next ask whether the nucleus contains anything which exists in duplicate in the fertilised egg but only as a single set in the separate germ-cells.

As stated, the cell nucleus of sperm and egg alike contains a substance called chromatin, which stains deeply. When a sperm and egg unite, the chromatin arranges itself in a number of filamentous structures, called *chromosomes* (Figs. 10 and 11), which mix with one another. Thus the egg and sperm each have half as many chromosomes as the fertilised egg. When the latter divide into two cells, the chromosomes split lengthwise,

Race, Reason and Rubbish

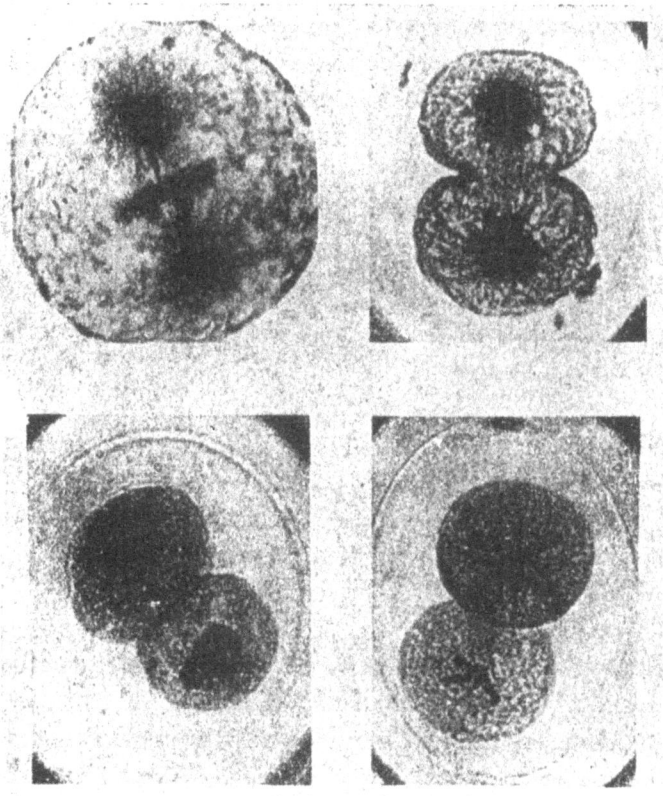

From ANIMAL BIOLOGY *by J. B. S. Haldane and Julian Huxley*

FIG. 10.—Photomicrographs of the first division of the egg of the Horse Thread Worm showing the four chromosomes

and both halves then move apart. In this process the two *centrospheres* (or *centrosomes*) play an essential part. Centrospheres are dot-like bodies from which radiating fibrous structures pass to the chromosomes. The latter are gathered up, each at its proper centrosphere; and in this way two new nuclei are rebuilt. Thereafter the cell body is constricted between the nuclei, so that two new cells are formed. When the two cells divide again history

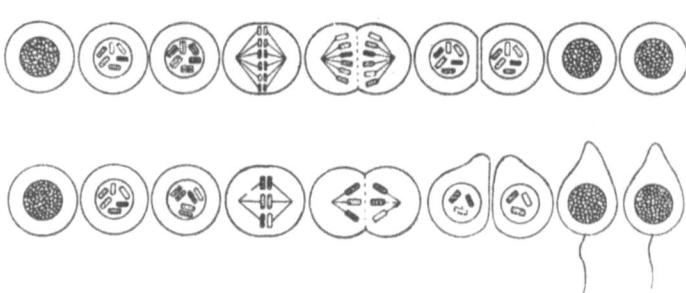

FIG. 11.—Diagrammatic representation of normal nuclear division in the growing tissues (including young germ-cells) above, reduction division before formation of sperm below

repeats itself. The chromosomes split lengthwise and the two halves each go where they belong.

In divisions which produce the vast assemblage of cells which constitute an adult being, the mechanism of division is consistently of this sort. When ripe eggs or sperm are formed, division takes place in another way. The chromosomes do not then split lengthwise but get together in pairs. One member of a chromosome pair goes to one pole and the second to another, and in this

Race, Reason and Rubbish

way the cells produced get only half the normal number of chromosomes. This is just what we should expect if the hereditary determinants are in the chromosomes. The fertilised egg and the individual which develops from it has double as many chromosomes as the ripe germ-cells. The two divisions immediately before the formation of the germ-cells, and the one by which the reduction of the chromosome number is brought about, are called the *reduction divisions*. Among some organisms it has been possible to see that a part of the chromosomes divide at the first of these divisions, while others pair and move off, each to its proper station. At the succeeding division those chromosomes which divided in the previous one pair in the same way.

Of course, it is a matter of chance whether a particular chromosome will or will not be in a germ-cell. If a determinant lies on a chromosome there should be an equally good prospect that it will be left in or left out of any particular germ-cell at the reduction division. In this connexion there is a trifling difference between the formation of the egg and sperm among higher organisms. The reduction division among the male germ-cells of animals produces only ripe sperm. In the reduction divisions of the egg, one large cell which will become the ripe egg, and small stunted cells which are thrown off, are formed. The latter are called *polar bodies*.

Any species has a definite number of chromosomes. For instance, the horse thread worm has only 4 chromosomes. The banana fly has 8, the fowl 74, the cat 38, the dog 78, and man has 48. So far as we know, the determinants are in the chromosomes. Hence we should try to get a more intimate knowledge of them. We should like to see them for ourselves, and to know about what sort of chemical composition the determinants have. Only lately have we begun to approach a solution of these questions.

The chromosomes are small structures. Even with strong microscopes it is difficult to see their finer points; but in a small insect—the banana fly (*Drosophila melanogaster*)—it has been discovered that cells which make up the salivary glands have relatively gigantic chromosomes in which it is relatively easy to make out details (see Figs. 19 and 20). These chromosomes are built up of darker and lighter strips with a somewhat variable appearance. From the standpoint of inheritance the banana fly belongs to the organisms most thoroughly investigated, and we can now identify from the appearance of the chromosome some of the determinants present in the banana fly from which the chromosomes are taken. To a certain extent we can thus foresee what characteristics the fly will have, and in this sense we can see the determinants themselves.

The conclusion that determinants occur in the chromosomes does not exclude the theoretical possibility that determinants may be found in other parts of the cell body, in the protoplasm of the egg, and the sperm

Race, Reason and Rubbish

respectively. In fact, we may take it for granted that the chemical constituents which make up the germ-cells among different species, for instance a fly or a human being, are of different kinds. Such chemical differences cannot be limited to the chromosomes; but if we do assume that there are chemical differences of the cell body among individuals of the same species, i.e. individuals which interbreed, such differences must be fixed entities. If so, there ought to be dilution in crossing of essentially the same sort as the old dilution theory implied; and such dilution would gradually lead to the production of a homogeneous substance. If there is a special constituent in one individual lacking among other ones of the same species, it would be diluted 50 % when crossing occurs, and after another generation reduced to a quarter, and so on. The result of this would be that the substance would be gradually reduced to a vanishingly small quantity among the offspring as it spreads itself out among the population. Only when determinants have the character of fixed entities can differences between individuals maintain themselves from generation to generation.

On this account it seems plausible to assume that hereditary differences between individuals within the same species are localised in the chromosomes alone. On the other hand, differences between species themselves are possibly localised in other parts of the germ-cells as well. Mendelism concerns itself primarily with hereditary differences within the species. Perhaps the old dilution theory may prove to be significant when species crosses are involved. With regard to animals, no proof for this belief is available. As regards plants, conditions are different in many respects. Among plants we find phenomena which signify that chemical differences of the cell body outside the nucleus may play their part in relation to inherited characteristics. For instance, we have reason to believe that white mottling of leaves sometimes depends on the composition of the cell body; but space does not permit further discussion of such problems.

CHAPTER II

DOMINANT inheritance, which has been explained already, is the simplest conceivable kind of hereditary process. If the determinant exists in an individual, so does the character. If the character itself is there, we can be sure that at least a single dose of the determinant is present, and that possibly it is there in duplicate. From a practical point of view the characteristic thing about this kind of inheritance is that if a person has the trait it is always present in at least one of his or her parents. Hence it never skips a generation. It can be traced backwards without a break in the ancestry.

We often hear people say that a particular characteristic cannot be hereditary because it has never, so far as we know, turned up in previous generations. Evidently, therefore, the popular conception of heredity is that when we inherit a character it must be traceable among some of our ancestors, or at least among some near relatives. In other words, people discuss human inheritance as if it were like inheriting money. We cannot inherit capital if our parents and all our relatives are poor. There must be wealth somewhere in the family, and it cannot remain concealed. Where hereditary characteristics of organisms are concerned, the state of affairs is quite otherwise. Nature deceives us even more than we ourselves deceive the income tax department.

Race, Reason and Rubbish

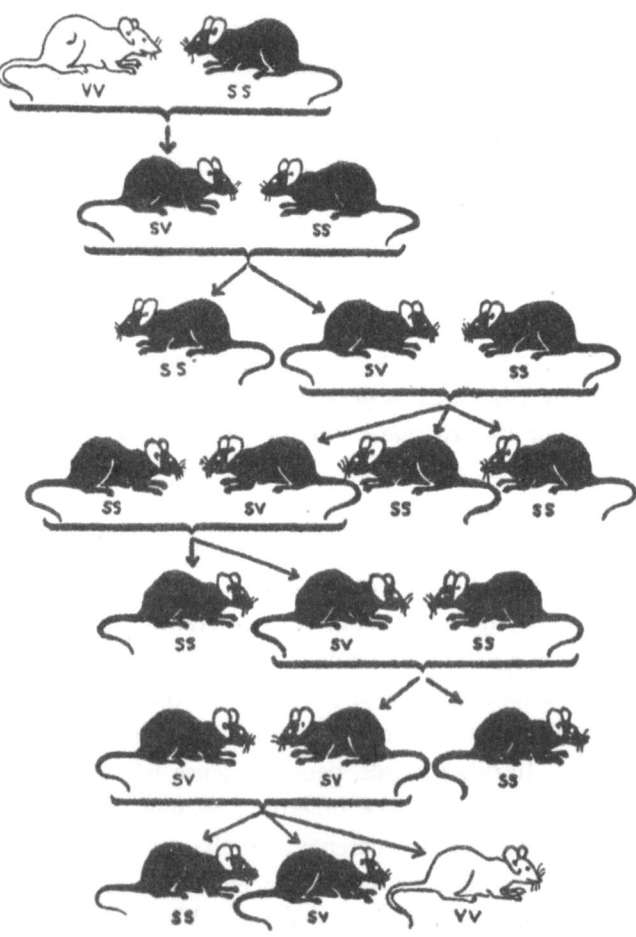

FIG. 12.—To illustrate that a recessive gene (V) may remain concealed for many generations of mating till two heterozygous individuals (SV) chance to mate (p. 49)

Race, Reason and Rubbish

We can inherit a character in spite of the fact that nobody in our family tree has had it.

Next to the preceding, the simplest sort of inheritance is the recessive type. The determinant, then, has no effect when a single dose of it is present. If, for this reason, *latent* in both parents, the determinants themselves may come out in duplicate and the character then appears among some of the offspring. In such cases we may often seek for it in vain among previous generations. To illustrate this we can take a fictitious example (Fig. 12). In the same laboratory we may have black and white mice which have separate cages. Suppose that owing to misadventure one of the white mice gets in among the black ones. This is soon discovered by the responsible attendant who puts the white mouse back in its proper place and hopes that no accident has happened. If mating has taken place in the meantime, a comparatively long period might elapse before noticeable results show themselves. When the white mouse mates with a black one, the offspring will be heterozygous, i.e. have only a single dose of the determinant for white. To all appearance they will not be different from other black mice, because black colour is dominant and white is recessive. The heterozygotes which turn up will probably mate with ordinary black ones which are homozygous. So we get only black mice in the next generation. This can go on for a long time until, by chance, two heterozygotes pair. On the whole, a quarter of the offspring will then

Race, Reason and Rubbish

be white. If a large collection of black mice is involved the probability that such pairing will take place is small. Perhaps it will be delayed for five or ten generations before the presence of the determinant is evident. This corresponds to several centuries when we are dealing with human beings.

It may therefore be a little troublesome to detect whether such a character is determined by heredity or not. It is even more difficult to take in the situation, when we want to disentangle how several characteristics are simultaneously inherited. An example from the behaviour of the banana fly (*Drosophila melanogaster*) will clarify this. The banana fly has been extensively used for experiments on inheritance. It is very easy to keep alive in the laboratory and quickly produces numerous progeny. The American geneticist Morgan and his collaborators have prosecuted the basic analysis of hereditary transmission in the banana fly. Without doubt it is now the most thoroughly investigated animal in the world from this standpoint. The wild type, as it occurs in nature, is brown with gray transparent wings and red eye-facets. Males and females have dissimilar marking at the hind end of the body, so that it is easy to distinguish the sexes. It lives on yeasts which flourish on fruit jelly and the like. So banana flies can be cultured in the laboratory in bottles at the bottom of which there is a little fruit jelly. Every twelve days a new generation comes out. Some of the flies are visibly distinguish-

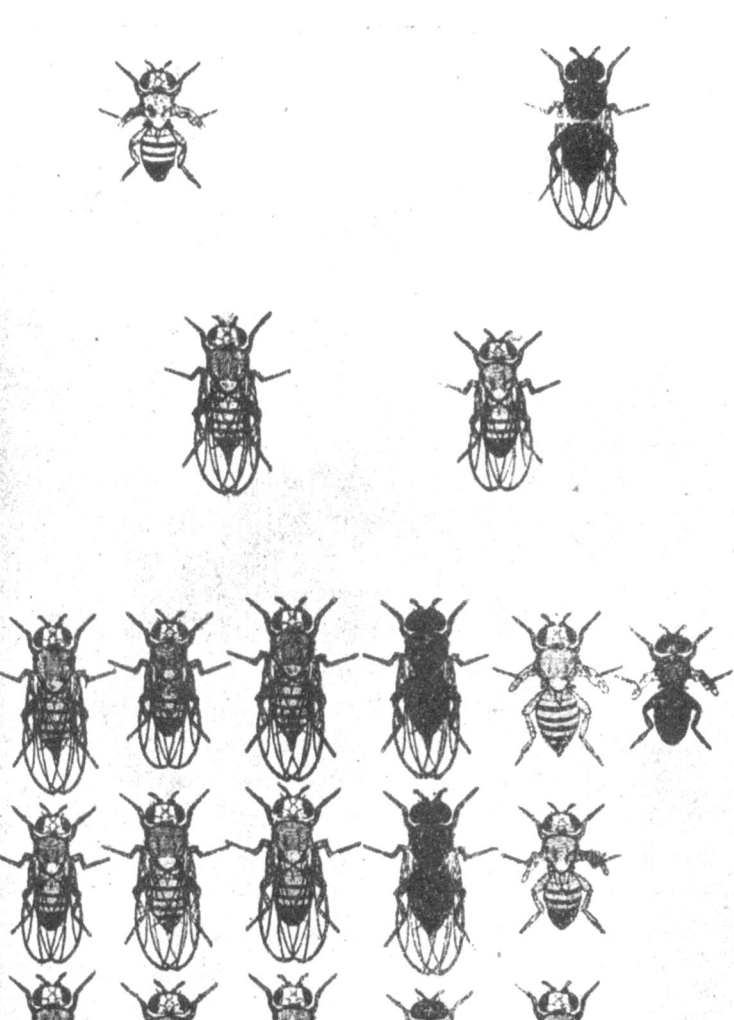

From EVOLUTION AND GENETICS *by T. H. Morgan*

FIG. 13.—Transmission of two recessive characters of the Banana Fly—mating of the recessive *ebony* mutant with the recessive mutant having *vestigial* wings

Race, Reason and Rubbish

able by hereditary characteristics, and these are of many different kinds.

If we cross a long-winged fly with a short-winged (*"vestigial"*) one the offspring are long-winged, and the determinant for short wings is thus suppressed, i.e. recessive. If we cross a gray fly with a black (*"ebony"*) one, the cross-bred flies are gray. The determinant for gray is dominant and the determinant for black is recessive. If we mate a short-winged gray fly (Fig. 13) with a long-winged black one, the first generation of crossbreds are both gray and long-winged. In other words they look like normal flies; but if we mate these crossbreds with one another the recessive characteristics come out again. From every sixteen individuals we get on the average nine which have long wings and are gray, thus having both dominant determinants and their associated characteristics. Three individuals are black and have long wings, i.e. flies which have one dominant and one recessive characteristic. There is the same proportion—three in sixteen—of gray flies with short wings—also combining one recessive with one dominant characteristic. In addition we get a single fly which is black and has dwarfed wings, thus showing both recessive characters because it has both recessive determinants in duplicate.

The relative frequencies of the four types are therefore 9 : 3 : 3 : 1. In order to get a better understanding of these proportions we must try to understand how determinants combine with one another. We will call the determinants for long wings *L* and for short wings *l*,

	GL	Gl	gL	gl
GL	GGLL gray long	GGLl gray long	GgLL gray long	GgLl gray long
Gl	GGLl gray long	GGll gray vestigial	GgLl gray long	Ggll gray vestigial
gL	GgLL gray long	GgLl gray long	ggLL ebony long	ggLl ebony long
gl	GgLl gray long	Ggll gray vestigial	ggLl ebony long	ggll ebony vestigial

FIG. 14.—Chessboard diagram to illustrate recombinations of two independent pairs of genes, i.e. genes located on different chromosomes

for gray colour *G* and for black *g*. The original parents then have the constitution *llGG* (short-winged gray) and *LLgg* (long-winged black). In the first generation the cross-breds get the constitution LlGg and must therefore be long-winged and gray. These flies produce germ-cells with the equipment *LG*, *Lg*, *lG* and *lg* respectively. They combine at random and the possibilities which are equally likely to happen can be got at with the help of the chess board diagram in Fig. 14.

For this purpose we simply write out the symbols for the germ-cells along the sides of a square, and put in the individual cells the letters which belong to the corresponding vertical and horizontal rows. In these cells we get sixteen possibilities we may expect for the make-up of the fertilised egg. Several of these combinations however have the same characteristics, and if we examine the symbols which stand for them, we get the different combinations in precisely the proportions given above, i.e. 9 : 3 : 3 : 1. Among the nine gray individuals with long wings, one alone is homozygous, and when crossed with its own kind produces only offspring with the same characteristic as its parents. The remaining eight gray flies with long wings have one or both of the recessive determinants. This can be shown by appropriate matings.

In order to check whether the result is in agreement with Mendel's Law we may follow the example of Mendel himself and deal with only one character at a time. Among the sixteen flies there are twelve with long and four with short wings, that is to say, one-quarter of the

Race, Reason and Rubbish

offspring have one recessive and three-quarters the corresponding dominant character. Similarly, if we look at the colours, four out of the sixteen are black. So one-quarter have the recessive trait. The possibility of getting either of the recessive traits is one-quarter. So $\frac{1}{4} \times \frac{1}{4} = \frac{1}{16}$ of the number of flies should get both of them together.

We can make calculations in the same way when still more pairs of characters are involved. Mendel himself carried out crosses between pea-plants with yellow seed-leaves, round seeds with grayish-brown husks, and plants with green cotyledons and wrinkled green seeds with white husks. The offspring of the first generation were peas of the yellow-round-grayish-brown type, i.e. with characteristics appropriate to all those dominant determinants. In the next generation, when cross-breds had been mated, the result was: 27 yellow-round-grayish-brown, 9 yellow-round-white, 9 yellow-wrinkled-grayish-brown, 9 green-round-grayish-brown, 3 yellow-wrinkled-white, 3 green-round-white, 3 green-wrinkled-grayish-brown, 1 green-wrinkled-white. Naturally, we can go further; and far more involved crosses have actually been investigated since Mendel's time. Fig. 15 exhibits a case in point. The picture shows a cross between two kinds of snapdragons (Fig. 15) which produce a hybrid with a prodigious number of different types.

If we calculate the chance that an individual with

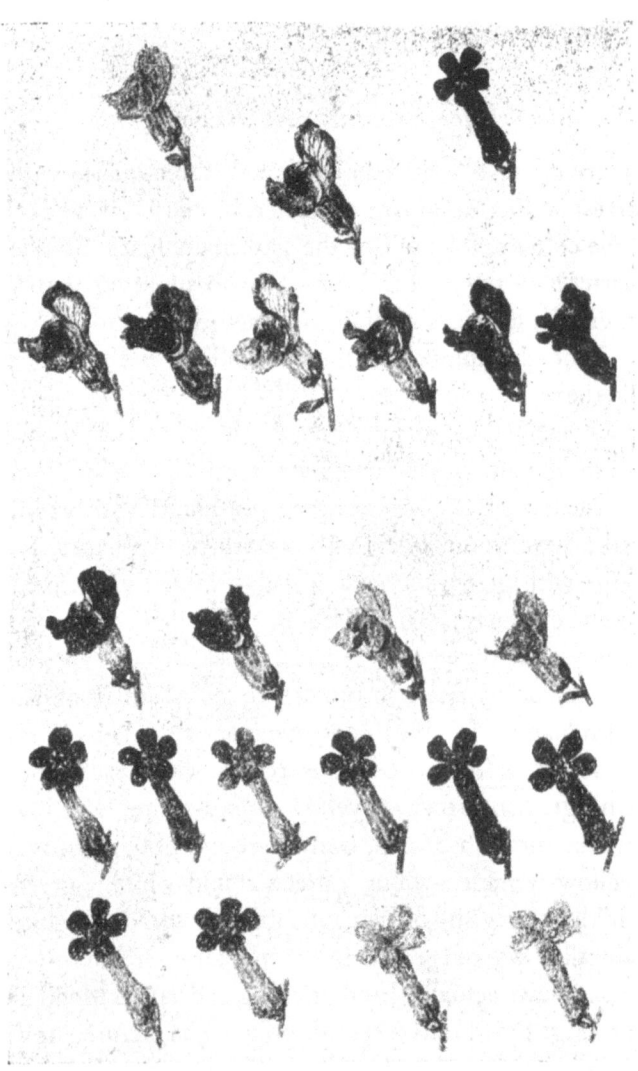

From EINFÜHRUNG IN DIE EXPERIMENTELE VERERBUNGSLEHRE *by Erwin Baur*

FIG. 15.—Cross between two races of Antirrhinum (snapdragons) distinguished by several genes affecting shape and colour of the flower

several characteristics will turn up, we find that the possibility is very small when it involves a considerable number of genes which combine freely. If both parents are heterozygotes for all the characteristics involved in a cross the probability that an individual will have any single recessive character is one-quarter. An individual with two recessive traits will turn up in one out of sixteen offspring. When it is a matter of three recessive characteristics, the required combination occurs once among sixty-four individuals. If we want a combination of four such characters we get it once among two hundred and fifty-six, and if ten of them are involved we have to hunt for one among a million—or to be more precise 1,048,576—individuals.

These figures make it clear that we have to face prodigious labour if we want to get an individual with a number of dominant determinants and one which also produces reliably homogeneous and constant progeny. We do not know which of the ones which display the desired characteristics, and to all appearances fulfil our requirements, have all the determinants in duplicate. In so far as the manifest characters are dominant, the majority of individuals have the determinants in single dosage. In so far as the combination we are out for includes recessive traits, we can deduce from the appearance of the individual if the requirements are satisfied. To this extent the problem is easier. However, we have to carry out a very large number of crosses and search among a very large number of individuals to get the type we want.

Stock or crop improvement is costly and time-consuming. For instance, if we want a strain of rye which has stiff straw, winter hardiness, early ripening, resistance to rust, large grain, and so forth, we face the prospect of going on a long while and undertaking heavy work. This is all the more so when the properties we are looking for are inherited in a more complicated way than we have so far dealt with. The truth is that, so far as grain improvement is concerned, we can soon reach useful results by applying scientific methods consistently. On the other hand, there is scarcely any limit to how long one can take to get more and more appropriate constant combinations by rule of thumb. Besides, the combinations we want are not the same for different countries; and even different parts of a country have different requirements, depending on soil fertility, climate, and the like.

For instance, we may need specially early ripening for varieties grown in more northerly regions where the summers are short. As is well known, Sweden and the United States have led the way in the work of improvement. It has been calculated (1936) that the work carried out in Svalöv increased the annual value of Swedish cereal crops by something like fifty to sixty million Swedish crowns. Obviously it is not easy to carry out a wholly exact estimation. We have to make a rough

Race, Reason and Rubbish

comparison between the value of the older types and the newly produced ones under *current* conditions of soil husbandry. Even if the reckoning is appreciably inaccurate there is no doubt that results obtained through applied scientific investigation in Svalöv alone repay many times the outlay of the State for scientific research and training in the universities or high schools. The allocation for this purpose in the annual budget of 1938–1939 was twenty-one million Swedish crowns.

If we look at scientific work exclusively from the economic standpoint, it is certainly good business for the community, and it would certainly pay in the long run, if larger allocations were set aside for it. Naturally, we cannot be certain that every increase of knowledge about heredity will produce a considerable gain. One cannot decide beforehand just what investigations will give results of practical value. To some extent scientific work is a lottery. If we have luck we get results, but results do not depend solely on luck, and in the long run we can usually be sure of getting results. In return for the sums which are being set aside during the next two decades, it is quite certain that discoveries of great importance will be made all over the world. If we educated more scientific workers and gave more of them opportunities for research work we could get the same results in a very much shorter time, perhaps in a couple of years.

Like America, Sweden has shown a comparatively

Race, Reason and Rubbish

intense interest in seed improvement, but stock improvement has only lately begun to attract attention. This is easy to understand because the prospect of getting results with stock are not so good as for crops. It costs more and takes a longer time. It is limited, among other things, by the costliness of the material. The time which elapses between successive generations is longer and the progeny are less numerous. Besides this, some characteristics which are specially interesting turn up only in one sex. This is true of egg-laying among fowls and milk production of cattle. With smaller domestic animals, however, the task is lighter than for larger ones.

Scientific methods applied to animal inheritance have only recently come into use, and in this field we have still to fight against old-fashioned beliefs. The following example is illuminating. Sheep bear twins nearly as often as they bear single lambs. Twin lambs have a somewhat higher mortality, and are somewhat smaller at birth, but they grow quickly and attain roughly the same body weight as lambs born singly when six to twelve months old. So it would be an advantage to get a strain of sheep which regularly bore twins. The sheep has normally two nipples; and it is an old tradition that sheep with supernumerary nipples bear twins more often than ordinary ones. Indeed, a similar belief has been put forward in more recent times, and has even been accepted as true, about human beings, i.e. women with supernumerary teats have been supposed to bear twins more

Race, Reason and Rubbish

often than others do. In his latter days Graham Bell, the discoverer of the telephone, got interested in sheep rearing and made an attempt to get a race which bore twins constantly. He proceeded from the prevalent belief about supernumerary nipples and chose sheep with such nipples for breeding. Gradually he succeeded in getting a flock which had six nipples. It transpired that the ewes did not bear twins more often than those of other flocks.

In other words, the belief in a connexion between the two characteristics, i.e. supernumerary nipples and the propensity to bear twins, proved to be false. A vast miscellany of such beliefs are held and have been held by animal breeders, and rules based on them still survive to-day. It is easy to understand how such opinions come about. When a farmer buys a cow, a horse, or the like, he cannot take the animal home for a time to test it. He has to decide at the sale. If he then makes a good purchase, and the beast proves to be a good one, he does not say that he has had luck but that he saw from the look of it that it was fine. Since it is a common human failing to consider that mishaps are due to luck and success to cleverness, he maintains that he went after some sign, some visible character. Indeed, even people who win in lottery prizes often pretend that they have some sort of second sight or that they felt constrained to buy a particular number among the ones they had to choose from. If the farmer fails in his purchase and gets a cow which does not give good milk, or something of

Race, Reason and Rubbish

the kind, he does not dare to say that he applied a false rule about the connexion between surface anatomy and milk production. In most cases he can be relied on to keep the secret of his misfortune to himself, and he tries to sell the beast as soon as he can. So we only get to hear about such rules when they appear to be successful.

Where domestic animals are concerned opinions about race-types are still taken seriously. Such beliefs are of the same general character as the one about the milking-sign referred to above. Thus it is a current belief that animals which conform to a certain breed-standard are necessarily better than animals which do not conform to certain rules laid down more or less arbitrarily. Now it is clear that such rules need not be consistently incorrect. There may well be a connexion between an animal's body-build and its performance. Animals which belong to a group with a particular average appearance may be, on the whole, better than animals which belong to another group. But if we want cows for the sake of their milk, and not because we like the look of them, the most reasonable thing to do is obviously to test whether they are really good milch-cows instead of trying to guess their power to produce milk from external appearances. If we go after appearances, we may get on to a wrong road and lose the way. At best we take a roundabout route and choose a trait which is comparatively irrelevant in the hope of getting a totally different one.

Race, Reason and Rubbish

We can put this in another way. Let us take it for granted that we are fond of a particular body-build, colour marking, or the like, because we want a "pure-bred" animal. Among the calves we sift out all which do not confirm to a standard appearance. After that we may eventually choose from those which are left over on the basis of their milking capacity. If we do not bother about appearances, but only think about the latter, we have more cows to choose from. It may well happen that a calf which has defective colour-pattern is just the one which has the best determinants for milk production. It is difficult enough to get hold of cows with good performance as milk-producers, i.e. which give plenty of milk with a high fat content, and we complicate the problem unnecessarily by demanding a particular appearance which may be quite insignificant.

Milking capacity is very difficult to judge, partly because it is inherited in a complicated way, and partly because nutrition and other circumstances have a powerful influence upon it. In this connexion resistance to disease, reproductive phenomena, and so forth, are also characteristics which have to be taken into account. So our problem is extremely difficult to solve. We should also remember that even if we succeed in producing animals which regularly transmit to their offspring a particular appearance, colour pattern, and so forth, it would be no guarantee for fixed transmission of other characteristics. To decree that the black colour of the

Race, Reason and Rubbish

Swedish lowland bull should not extend below the knee or hamstrings is therefore pointless. If we are primarily concerned with meat production, as happens in some countries, it is not so unreasonable to be guided by external appearance, but naturally one should not bother about colouring, the set of the tail, shape of horns, and such things.

To illustrate the part which erroneous ideas about the outside of an animal have played in the work of breeding, particulars of the prize standards of the Twenty-third Public Agricultural Conference at Stockholm in 1930 are cited below. Among the cattle, bulls were awarded prizes partly for ancestry up to a maximum of fifty points, and partly for external traits up to a maximum of fifty points. The external traits taken into consideration were:—

Head, horns, neck	maximum 5 points
Chest, shoulder, withers	,, 5 ,,
Back and loins	,, 5 ,,
Rump	,, 5 ,,
Set of tail	,, 5 ,,
Thigh and groins	,, 5 ,,
Colour markings	,, 5 ,,
Legs and tread	,, 5 ,,
General impression	,, 10 ,,
	Total .. 50 ,,

Of itself, as is quite clear, it is completely irrelevant

what colour and marking or what sort of tail the creature has. In this list there is nothing which bears in any way directly upon the animal's productivity. However, yield does come into play for points given on account of its ancestry. Here the requirements are as follows:—

Father's pedigree	maximum 10 points
Mother's pedigree	,, 10 ,,
Mother's milk yield	,, 10 ,,
Paternal and maternal grandmother's milk yield	,, 10 ,,
General impression	,, 10 ,,
Total	50 points

In the total of one hundred points which could be gained by a bull his own progeny have no place. The mother's and father's and grandmother's progeny account for a maximum of twenty points altogether. Since 1930 regulations have changed to some extent, and are now more sensible. In this connexion we need not go further into the complicated question of methods for prize allocation and stock improvement. Space forbids it, and it is not the author's intention to criticise. The schedule is cited merely to give a practical illustration of how false beliefs about problems of heredity continue to have a vogue.

When human beings are the topic of discussion we meet just the same erroneous notions. We all know of people who talk about race types, as if the appearance

Race, Reason and Rubbish

of an individual and his descent from individuals of a particular shape or colour is of more importance than his social capabilities. If we want a worthy and intelligent person for a particular post, the only guarantee we have of getting one is to test, as best we can, whether candidates have the characteristics required. As a rule we do not take anyone on for a job because we want a person with a particular appearance. To be biased by race type is just as perverse where human beings are concerned as it is when we are dealing with cows. If we want cows for their milk the reasonable thing to do is to test if a cow possesses this merit. These questions will be discussed more fully in another context. Here it is enough to emphasise how necessary it is to apply scientific methods, and to meet the expenses incurred in using such methods as are employed for stock improvement of domestic animals.

Perhaps some of us will next ask whether there is any connexion between inherited characters so far as their transmission is concerned. In reality we know quite a lot nowadays about different kinds of connexions between inherited characters. The simplest type is when two characters depend on one and the same determinant. If one character is then present the other is also and always present. An example of such characteristics is illustrated in Fig. 16. It has been found that one determinant which is responsible for dwarf wings in banana flies also leads to the presence of two small bristles in

Race, Reason and Rubbish

the middle of the trunk. So also pea-strains with coloured flowers always have seed pellicles which are coloured.

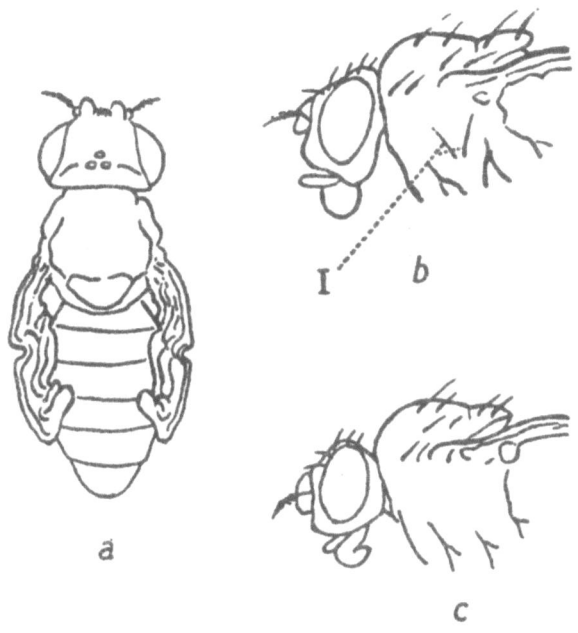

FIG. 16.—Multiple effects due to a single gene

We recognise a gene by the most striking effect it produces; but it is wrong to think that each gene has one effect and only one. Thus the gene which produces vestigial wings (*a*) when present in duplicate in the cells of the banana fly, also produces two extra bristles I (*b*) not present in the wild type (*c*)

A clear-cut connexion of this kind is not very usual, but it has been possible to elucidate one which is more common in the banana fly and several species of plants.

Race, Reason and Rubbish

In discussing what happens when the material basis of two characters lies on the same chromosome, we shall henceforth use the more modern and shorter word *gene* instead of the older and longer word determinant; and we must now retrace our steps to what has been said about the behaviour of the chromosomes themselves. How a particular chromosome comes to be present in a germ-cell depends on chance. The chromosomes of a pair come apart at the reduction division, and if two genes lie on chromosomes belonging to different pairs it is a matter of chance if they remain together again or get separated. In any species there is only a limited number of chromosomes and consequently the genes must be distributed in blocks which are equal in number to the chromosomes themselves. So we might expect that genes of the same group, i.e. genes which belong to one and the same chromosome, would keep company.

The banana fly has four pairs (Fig. 17) of chromosomes and one would expect that the genes of the banana fly would stick together in a corresponding number of groups. To a certain extent this is true; but it has been found that even when two genes do belong to the same group, i.e. when they lie on the same chromosome, they do not always stick together on that account. This conflicts with what we have been led to expect. So there must be some special arrangement which ensures that genes can sometimes get separated.

When genes keep company because they lie on the

Race, Reason and Rubbish

same chromosome they are said to be *linked*. Our problem is, therefore, how linkage is sometimes broken (see Figs. 34 and 35). The American geneticist T. H. Morgan and his associates have put forward a theory to

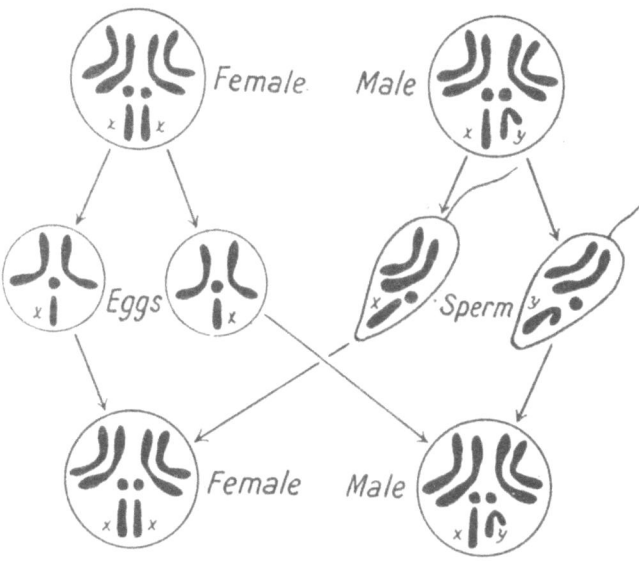

FIG. 17.—Sex determination in the Banana fly. Chromosomes in normal cell division of male (right) and female (left) top row, and gametes in middle row

explain this occurrence. They suppose that chromosomes of the same pair at the reduction division do not lie parallel but are twisted together. When they separate afterwards it sometimes happens that they break, and that a region of one unites with the neighbouring region of another, as shown in Fig. 18. This phenomenon is

Race, Reason and Rubbish

called *crossing-over*. It has now been possible to see that chromosomes really do behave in this way.

We can use the figures which result from linkage to calculate the distances which separate genes on a chromosome. The prospect that two chromosomes will separate if the genes lie near each other are small, and large if they are far apart. It is clear that the greater the distance

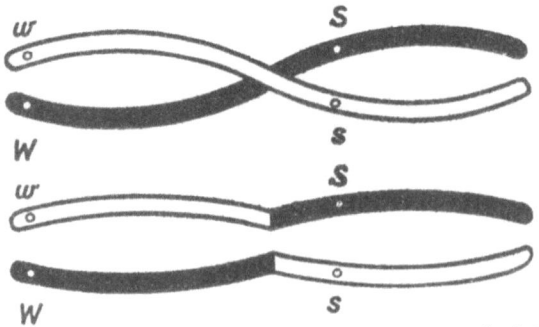

From HEREDITY *by A. F. Shull*

FIG. 18.—Diagram to show crossing-over of genes owing to twisting of chromosome when the pairs separate at the reduction division

between them the greater will be the probability that crossing-over between two genes will take place. If two genes a and b get apart in 3 %, and b from another gene c in 2 % of the reduction divisions, the percentage separation for a and c should be $3 + 2 = 5$ %, when a and c lie on opposite sides of b. If c lies between b and a, c and a should come apart in only $3 - 2 = 1$ %. We can thus calculate in advance what proportions we ought to get if we have found out. If the genes lie far

Race, Reason and Rubbish

from one another, i.e. each at different ends of a chromosome, crossing-over seems to occur to a less extent than we should expect on the basis of such calculation. This depends on the fact that double crossing-over between genes happens in some cases, so that they come to lie on the same chromosomes.

At first the theory of crossing-over met with strong opposition. This is not surprising. One has to put up with fair criticism of any new theory, and there is no place for a scientific theory before it has been seriously discussed. There must be relatively strong support before it is accepted in the sense that one can build further upon it. In science one always tries to sift fact and theoretical opinions about causal connexions which must always be subject to revision, however reasonable or congenial they appear to be; but in different circumstances criticism is not equally severe. As a rule we try to reach explanations that are as simple as possible. In most circumstances we prefer a simpler explanation to one which is more involved, but there is also a disinclination among scientific men to adopt theories which look too simple. We have the feeling that problems which touch on the nature of life and the like, and ones which are more or less fundamental, must lead to complicated explanations. So we hesitate to accept a very simple one and may sometimes be a little hostile to it.

This cautious attitude is both natural and understandable. So far as the theory of crossing-over is concerned, there was perhaps a feeling that the picture of genes as entities which lie like beads on a thread is a

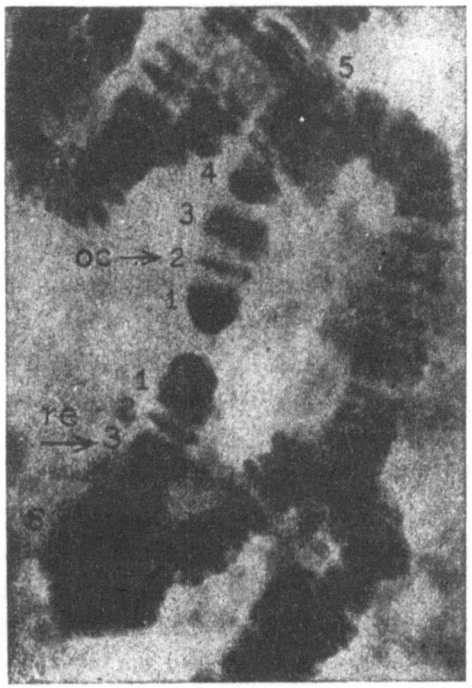

From JOURNAL OF HEREDITY

FIG. 19.—Photomicrograph of part of a chromosome of the salivary gland of the Banana fly, showing dark stripes to each of which identified genes can now be assigned

little too simple. We do not like to imagine that genes which determine whether one of us is a genius or another an idiot are small particles in single file like

Race, Reason and Rubbish

convicts in a chain gang. All the same, we have now got satisfactory grounds for believing that Morgan's theory agrees with facts so closely that we no longer seriously doubt its validity. Indeed Morgan, who has been awarded the Nobel prize for his work, is the first geneticist who has held it up to date.

Thanks to the theory of crossing-over it has been possible to chart the chromosomes of the banana fly, and subsequently of some plant species, so that we now

From JOURNAL OF HEREDITY

FIG. 20.—Part of one of the chromosomes of the salivary gland of the Banana fly—the dark stripes correspond to the situation of genes which are known

know in what way many hundreds of genes are arranged. At present we know this about some four hundred genes of the banana fly. It has even been possible to identify the position of genes in the finest details (Figs. 19 and 20) of the structure of the chromosomes. So it is now possible to study the chromosomes of the banana fly and say what characteristics an individual fly will have. If we show a microscopic preparation to a specialist on the banana fly, he may be able to tell whether the fly itself had dwarf wings or not, whether there was any modification of the eye colour or body markings, and so on, without seeing it.

CHAPTER III

So far we have only talked about the simplest types of transmission. There are also others which are more complicated, despite the fact that they can be interpreted by means of the simple Mendelian laws. So-called *polymeric* characters illustrate a special complication of this sort. Such characters depend on a number of genes which have a complementary or antagonistic influence. If a majority of the positive type is present, the character so determined will be displayed to a greater extent than if there is a majority of negative ones.

The height of a human being probably belongs to the class of characters which depend on several genes. The genes reinforce one another. The greater the number of those which bring about increase of height, the greater will the height of the individual be. Such transmission has been thoroughly demonstrated and analysed for the black colour of oats, size of maize cobs, etc. Egg laying and milk output in animals probably depend on similar cumulative genes. It is obvious that if we want to bring about an improvement with respect to such characters the difficulties are considerably greater than when it is a matter of characteristics inherited in a straightforward way.

Another more intricate type of inheritance may also occur when a character depends on several genes coming

Race, Reason and Rubbish

together. In simple recessive inheritance we have to get two genes of a pair together in order to show up the recessive character. In double recessive transmission we have to have two recessive genes in duplicate, i.e. four genes must come together in the same individual to produce the recessive character. When only one pair of genes is involved, we speak of monohybrid inheritance, and when two are significant for the production of a character, we speak of dihybrid inheritance. When this is so, we can naturally get different kinds of combinations of the recessive and dominant genes. A particular character may depend on several recessive or several dominant genes which must all be present before it will show up.

In the course of genetic research there has naturally been more success in finding out about characters inherited in a simple, as opposed to a complicated, way. The more complicated types have been dealt with best among plants, and are less well known among animals. So far as man in particular is concerned, we have only been able to interpret the simplest phenomena till now. Meanwhile we may suppose that the majority of characters are more complex. The difficulties which beset further analysis of such characters are very much greater. Since scientific work in this field has to be continued for several decades, it is not surprising that our knowledge is defective, though we can certainly expect to get results both of theoretical and of practical value in

Race, Reason and Rubbish

future. Meanwhile, work already carried out shows that the problems are more complicated than we first imagined. Anticipations which we used to entertain about getting practical results quickly were exaggerated.

In the ensuing chapter we shall try to see why this is. If we ought not to be excessively optimistic there is no reason for pessimism. Even in the quite short interval since orderly scientific work began we have gained the right to make definite assertions about some issues. The last hundred years have been called the Age of Electricity. It is possible that we shall one day call the twentieth century the Age of Biotechnics. If so, it will certainly be due to the discoveries which have been and will be made through research about heredity. It is possible that we shall call the succeeding century the century of Social Science. In the meantime we can at least hope that scientific methods and opinions rooted in scientific knowledge will day by day play a more significant role in our common life. The time for being consistently cheerful has not yet come.

Among the more complex kinds of inheritance we should now take stock of two special types. It does not always happen that genes located in different pairs of chromosomes reinforce one another. Just as we can distinguish (p. 38) between intermediate inheritance, extramediate inheritance and dominance when we confine ourselves to a *single* pair of genes, we can also distinguish between three ways in which genes of

Race, Reason and Rubbish

different pairs contribute to the characteristics of an individual. The first type, corresponding to intermediate inheritance with respect to one pair, is illustrated by polymeric transmission of the black colour of oats, already mentioned in the second paragraph on p. 47. Usually, if not always, it implies that inheritance for any single pair is of the intermediate type. If *a* and *b* contribute equally to a character of this class, a plant with the constitution AaBb is much the same as a plant with the constitution AAbb or aaBB. A phenomenon analogous to extramediate inheritance involving two pairs of genes is well illustrated by the inheritance of comb shape in fowls. In contradistinction to the common *single comb* of the Mediterranean breeds (e.g. Leghorns) fowls may have the types (Fig. 21) called *pea-comb* (e.g. in Partridge Cochins) or *rose-comb* (e.g. Hamburgs, Wyandottes, and Sebrights). *Pea* and *Rose* are each dominant to *Single* and depend on single genes. When fowls with pea combs are mated with fowls with rose combs the offspring have a new type of comb called *walnut*. This is characteristic of the Malay breeds, and is in no sense intermediate between the parental types.

A third possibility analogous to simple recessive inheritance gives rise to peculiar results. It occurs when each of two recessive genes *a* and *b* may have the same effect in duplicate but do not reinforce one another. Thus individuals with the constitution *aa* and *bb* may look alike and both show what appears to be the same

Race, Reason and Rubbish

recessive character, just as individuals with the constitution *Aa* or *Bb* show the same dominant character; but individuals with the constitution ab (i.e. *AaBb*) are not

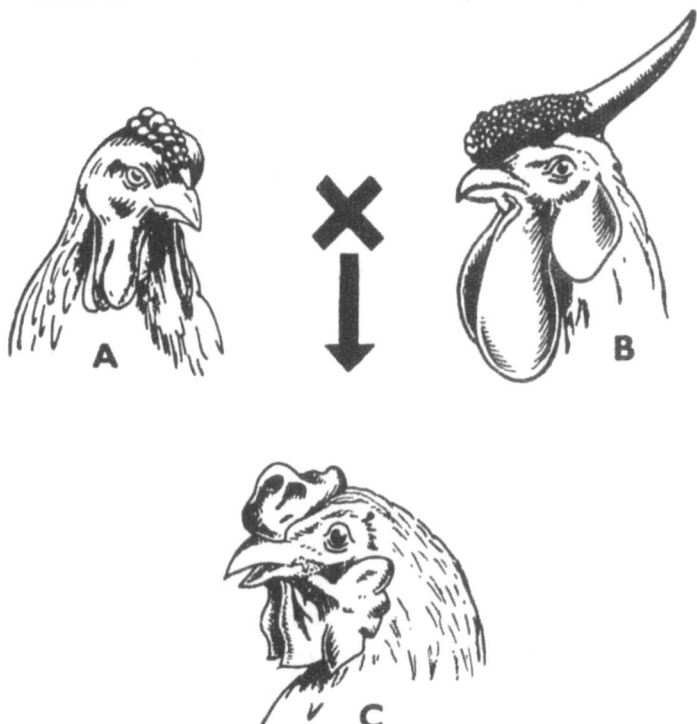

FIG. 21.—Inheritance of Comb shape in Poultry

like individuals with the constitution *aa* (i.e. *aaBB* or aaBb) and *bb* (i.e. AAbb or Aabb). They are of the dominant type indistinguishable from individuals with the constitution AaBB, AABb, or AABB.

Race, Reason and Rubbish

An example of this kind turns up in crosses involving coat colour of animals; for instance, between pure-bred strains of white rabbits which always bear white offspring when kept by themselves, but have dark ones when an individual of one strain is mated with one of another. Though close inspection reveals slight differences, e.g. that one strain has pink eyes and the other has pigmented ones, the coat colour of the two is indistinguishable. The white hair of each strain depends on a particular recessive gene, which we may call *a* in one strain and *b* in another. When we are discussing an outcross to a coloured strain involving the former we can represent the dominant type by *AA* (homozygous) or *Aa* (heterozygous), and when we are speaking of an outcross involving the second we can represent the dominants by *BB* or *Bb*. To take all the relevant facts into consideration we should therefore represent pure-bred whites of one strain by *aaBB* and pure-bred whites of another by *AAbb*. The offspring of a cross between them must have a single dose of both recessive genes *a* and *b*, i.e. they will be *AaBb*. The fact that they are coloured shows that although the presence of *aa* has the same general effect as the presence of *bb*, the combination *ab* has no visible result, i.e. that *a* and *b* do not reinforce one another.

This can be proved by subsequent mating of the coloured cross-breds. According to the explanation given on page 54, nine out of sixteen offspring of these should have neither recessive gene in duplicate, if they are independently assorted, nor, in other words, if they are located on different chromosomes. This means that seven out of sixteen have either or *both* in duplicate and are white. Experiment shows that this is roughly the proportion in which whites turn up. Among the whole progeny one out of sixteen or one in seven of the white progeny should have both *a* and *b* in duplicate, and should therefore have white offspring when mated with white rabbits belonging to either of the original white stocks. Such individuals can be recovered in experiments carried out in this way, and furnish additional confirmation that this explanation is the correct one.

Another interesting type of result which occurs in crosses between animals is the following (Fig. 22). There are mice which are yellow and always give a small proportion of yellow and a larger proportion of gray offspring when mated. This means that the yellow colour is dominant and that all such yellow mice are heterozygous with one gene for yellow and another for gray hair. If their offspring are mated with one another we should expect to get some yellow ones which regularly produce yellow offspring. We should also expect that one third of the yellow ones would be homozygotes; and if we persisted in mating them we should quite often—in one case out of nine (i.e. $\frac{1}{3} \times \frac{1}{3}$)—make matings between homozygotes which produce only yellow progeny. As it happens, we never do get this result: In other words, there are no homozygotes among the offspring. All yellow mice are heterozygous.

If we examine the numerical proportions we get when we cross yellow mice, we find that one-third of the offspring are gray. In crosses between heterozygotes we ought to get one-quarter gray individuals, if gray

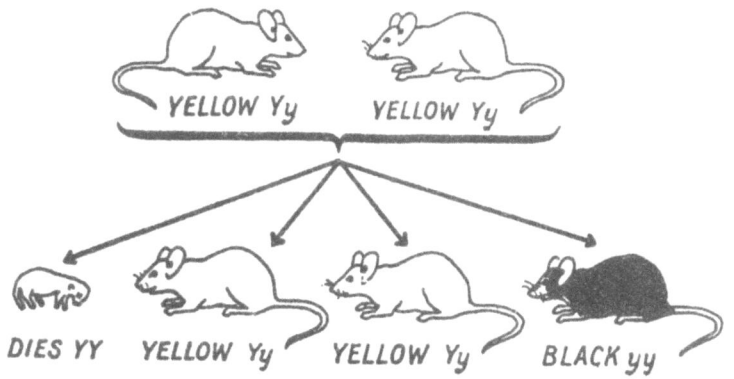

Fig. 22.—Diagram of the inheritance of the Lethal yellow coat colour of mice

is recessive, but if we exclude homozygotes, we should have two yellow heterozygotes for every single gray homozygote, i.e. just the proportion (one-third gray and two-thirds yellow) which we do get. For some reason or another, it looks as if two gametes which both have the gene for yellow cannot unite or, if they do, produce offspring which do not survive. In fact we have two possibilities. One is that such gametes are generally incapable of uniting. The other is that they actually do so but that the individual which then develops dies before birth. The second alternative can be shown to be the correct one.

If we examine such mice during pregnancy we find that a quarter of the average litter consists of embryos which are imperfect and die. So the gene for yellow colour in duplicate brings about death. Such genes are called *lethal* or mortality genes. Many such genes which bring about defective viability are now known. They are of practical importance among domestic animals because they may lead to miscarriage or stillbirths. One known case occurs among Swedish cattle. Two highly esteemed bull strains, *Prince Adolf* and *Galens*, have been shown to have had lethal genes. In this way a large stock of Swedish lowland cattle has been infected with them.

Another peculiar process which proves to be of practical importance, but advantageously, depends on *chromosome doubling*. In exceptional circumstances it may happen that the chromosomes do not go to opposite poles as at ordinary reduction divisions when germ-cells are formed. There is a half-hearted attempt at nuclear division, and the cells which are formed may afterwards get fertilised. In this way we get an egg which has twice as many chromosomes as an ordinary one. Much the same thing can happen with sperm, and in this way we may get individuals with half as many, or it may be twice and even several times as many, chromosomes as are characteristic of ordinary cells in ordinary individuals.

Such increase of the chromosome number generally leads to increased bulk of the adult itself. In fact, we get giants. That such individuals are larger than the average depends directly on the fact that the cells from which they are built up are themselves larger than ordinary ones. Naturally, the nucleus must be larger than usual because it contains more chromosomes, but the greater size of the nucleus seems also to produce an enlargement of other parts of the cell. Chromosome doubling occurs particularly in plants, and now plays an important part in crop improvement. It is possible to make use of plants which turn up accidentally with a double set of chromosomes (Fig. 23), and special methods have been worked out for promoting chromosome-doubling. We can take advantage of the influence of abnormal temperatures, particularly exposure to cold, and also of the effect of a group of poisons which specially affect cell division, e.g. chloral hydrate, nicotine, and quinine sulphate. One drug which specifically affects cell partition is *colchicin*, used as a medicine for gout. In appropriate dosage it stops cell division at a particular state, a phenomenon of great interest from the biological point of view.

Chromosome doubling does not merely lead to a change in the size of a plant. It can also produce changes

Race, Reason and Rubbish

of other characteristics. New hereditable properties which depend on new genes are called *gene mutations*. Those which depend on chromosome doubling or other sorts of chromosome redistribution in gametes are

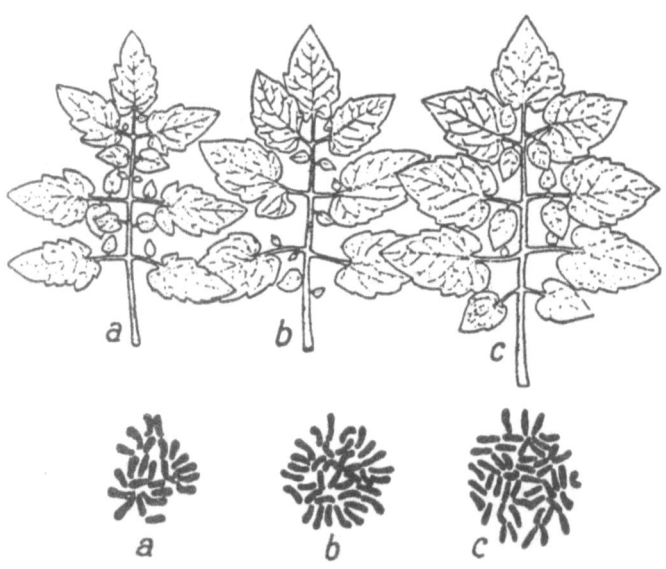

From AN INTRODUCTION TO MODERN GENETICS *by C. H. Waddington*

FIG. 23.—Chromosome Mutations of Tomatoes. The three varieties are distinguished by possessing 24, 36 and 48 chromosomes

called *chromosome mutations*. The term mutation was introduced by de Vries, a Dutch scientist. De Vries believed that he had discovered in the Evening primrose (*Oenothera biennis*), a cultivated plant (Fig. 24) which also grows as a garden escape in Britain, changes that occurred among the offspring at a single step, suddenly and without any detectable reason. He also thought that this was due to the production of new genes. In reality

From EINFUERUNG IN DIE VERERBUNGSWISSENSCHAFT *by R. Goldschmidt*

FIG. 24.—The Evening Primrose (*Oenothera Lamarkiana*)

it has turned out that this somewhat isolated emergence of new types depends on other processes of inheritance, i.e. on complicated types of chromosome splitting, and

Race, Reason and Rubbish

that the new types are not *gene mutations*. The phenomenon to which de Vries gave a name, and the one in which his researches awakened interest, exists in nature as well as in the laboratory. For some unknown reason chemical changes do occur in the hereditary materials. As the result of a mistake, de Vries gave an impetus to a large number of investigations, and he was in the right although the reasons on which he based his views were not correct. Even in scientific work good luck and bad luck play a role which is not negligible, and this gives scientific research a charm of its own. A scientist who is badly equipped, one who, humanly speaking, would not be expected to discover anything noteworthy, may thus have the good fortune to get results of outstanding significance.

If new gene mutations turn up we should expect that the same changes would do so twice or more times. A gene mutation which brings about a change of eye colour in the banana fly ought not to occur only once, but should be observable several times. Naturally, the likelihood that a gene-change will occur twice is greater when the total number of genes is small than when the total number of genes is great. From this point of view we can actually make a rough estimate of how many genes the chromosomes of the banana fly contain on the basis of the ration of single to repeated mutations. Such a calculation has been made. In round figures the result is that the fly must have at least one thousand

Race, Reason and Rubbish

eight hundred genes. This figure is decidedly a rough one, but, all the same, it gives us an approximate statement of the order of magnitude involved.

As emphasised, mutations are comparatively infrequent phenomena. They occur in animals as well as in plants, and must therefore occur in man also. We have also good reason to believe that more or less rare inherited diseases of different sorts have arisen by mutations in the first place. We have also good reasons for believing that mutations with similar effects have turned up several times in the population, because sicknesses which appear to be the same are inherited in different ways in different families—in one family sex-linked (page 68) and in another as an ordinary recessive condition. Further investigation shows that such diseases, though broadly similar, are distinguished by small differences.

Gene mutations are specially interesting, when we take a long view of problems, and they are most significant when we want to explain how organisms evolve from lower to higher forms, that is to say evolution from relatively simply constructed organisms to more complex species. In part we can explain evolution by assuming a new combination of genes. Chromosome doubling (*polyploidy*) and other irregularities of chromosome distribution (Fig. 25) also plays a role (Figs. 26 and 23), which is by no means insignificant. Many of our strains of cereals and the like have probably arisen

Race, Reason and Rubbish

in this way. Most of all in this connexion we must take into account the possibility of accidental hybridisation between different species belonging to the same genus, so that results of polyploidy are combined. Unlike species in a genus often display a graded series of chromosomes. For instance, one species has five pairs, another ten, a third twenty, a fourth twenty-five, a fifth

FIG. 25.—A chromosome mutation in the Banana fly. A segment of one member of one pair is attached to a member of another pair equally mated in cells of the normal fly as seen on the right

and seventh respectively thirty and ninety. These are actually figures from the plant genus *Senecio*.

Even if new combinations of genes can explain the origin of many new species in one way or another, it is scarcely likely that this is the whole truth of the matter. Nevertheless there are men of science who assert that such a mechanism is completely adequate to explain the evolution of the diversity of species found on the earth. According to this view all genes were present in the beginning, and the way in which they are combined

From THE CAUSES OF EVOLUTION *by J. B. S. Haldane*

FIG. 26.—The normal hybrid (left below) *Primula kewensis* produced by cross pollination of *Primula foribunda* (left above) and *Primula verticellata* (right above) is sterile; but a chromosome mutant of *Primula kewensis* with double the normal number of chromosomes is a new fertile species

in increasingly intricate collections is all that distinguishes the higher from lower organisms. That people are not disposed to assume the origin of really new genes is explicable for several reasons. By avoiding this assumption we relegate the problem of how genes come into existence to an unknown and mysterious beginning of time, when organisms first began to appear on the earth. A further reason for doing so is that it is difficult to imagine how completely new genes arise.

A new gene must correspond to a particle which produces new chemical reactions, and hence new characteristics, when introduced into the gene complex. It must at least have a material basis of some sort. The author's view is that we have to choose between two possibilities if we want to discuss this issue. One is that new chemical constituents can get into the sex cells from the outside world. This is difficult to imagine and is not very probable. It is scarcely credible that particles which can make their way into the germ-cells have the power to bring about the production of new genes. More likely perhaps is a second possibility that the particles which give rise to new genes come from the egg or the sperm itself, and more particularly from its own gene complex.

We can imagine that a group of atoms get detached from the gene complex. This may result in the destruction of a gene, and we get mutation by loss. The gene equipment of the individual is changed in a negative way, that is to say, it now has a simpler structure. Such a chemical change can naturally bring about the appearance of something new which produces the impression of a positive characteristic in the organism as a whole. We can also go a step further and imagine that the group of atoms split off gets attached to another molecule or to a different position in a molecule which constitutes a gene. A new gene with a more complex structure than had previously existed may perhaps turn up in this way. If the newly attached group of atoms comes from another gene, the cell concerned may have lost one gene in the process of gaining a new one. Since we can never get to a more complicated gene equipment by merely losing as much as we gain, the loss must be covered in some way. This may happen, owing to reduction and the new combinations which sexual reproduction involves. Owing to the existence of the reduction division a chromosome which has suffered loss may be rejected. When a normal equipment of chromosomes is contributed by the other gamete, we get an individual which has lost no gene but possesses a new one in single dosage.

Among lower organisms reproduction (Fig. 27) may come about through budding. This may be said to imply that some cell of the adult reverts to the embryonic condition and begins to develop into a new being. The latter remains attached to the parental body but separates when it reaches a certain size. If a cell which gives rise

Race, Reason and Rubbish

to a bud destined to develop into a new individual gets a new gene, it may also have suffered a loss at the same time. In reproduction of this sort, called *vegetative* reproduction, loss of genes can never be covered. On this account sexual reproduction may be a necessary, as

Fig. 27.— Vegetative reproduction of the Fresh Water Polyp *Hydra*. The organism consists of a contractile trunk attached at one end to a solid object. The mouth is at the free end surrounded by grasping tentacles

well as advantageous, condition for the evolution of more complex organisms. In lower animals vegetative and sexual reproduction, or at least some mechanism which more or less corresponds to sexual reproduction, often occur in the same individual.

Space does not permit more detailed treatment of the problem. What must be emphasised is that, if the

Race, Reason and Rubbish

schematic view put forward above is a correct one, it furnishes an explanation of why sexual reproduction has prevailed and is everywhere found among higher organisms. At the same time there is nothing to prevent vegetative reproduction at a comparatively high level in the world of organic life. To be sure, the mechanism suggested merely implies a little modification of the phenomenon of recombination which, as we certainly know, has played some part in evolution. When we talk of new combinations, we may concede that only a portion of the genes can be recombined. Throughout the ages the wave of living matter sweeps tirelessly over the earth we live in. Living matter becomes more and more complex with ever new combinations of ingredients, as small drops of spray unite ceaselessly with one another. Like beads of foam cast up by the wave, individual lives glisten for a moment in the sunlight and fade away. Living beings die and disappear, but the wave of life rolls onwards. Organic matter becomes more intricate and passes into what we call higher forms. For the twinkling of an eye our human family mounts the crest, but we do not know whence came the wave which carries us, or whither it is going.

CHAPTER IV

In former times people have put forward a large number of explanations for the fact that some individuals are male and others female. They have tried to prove that the mother's age must play an essential role. They have also tried to put over the belief that the father's age is important, and have even believed that the age of the egg could exert a decisive influence. Some have striven to show that the nature of the diet may affect the sex of the developing individual, and have tried to modify the sex of the foetus by letting mothers feed on particular food constituent during pregnancy. Through efforts of this kind it has gradually become clear that sex is fixed at a very early stage in the development of the embryo.

Discoveries made in research on inheritance have now made it clear that sex is really determined at fertilisation. It has been found that males of some species have one chromosome less than females. For example, this is true among insects such as grasshoppers. In one species of grasshopper there are thirty-two chromosomes in every cell of the female body, but only thirty-one in males. Among such grasshoppers one chromosome pair of the female is decidedly larger than the remaining ones and is therefore easy to recognise. With the microscope we can see that females have two

Race, Reason and Rubbish

such large chromosomes without any difficulty. At the reduction division, when the gametes are formed, the normal number of chromosomes is diminished by half. So the females produce eggs which have sixteen chromosomes. At the reduction division in the male two sorts of cells are produced. The extra chromosome accompanies one lot of gametes which thus get sixteen chromosomes. Others are formed without the extra chromosome, and these have only fifteen. The extra one is called the sex chromosome or X-chromosome.

Thus there are really two different sorts of sperm. If an egg is fertilised by one which has the sex chromosome, and therefore has sixteen altogether, the fertilised egg gets thirty-two. The individual which develops from the latter has two sex chromosomes in all its cells. It therefore belongs to the female sex. On the other hand, if fertilisation is brought about by a sperm which has only fifteen chromosomes, the new individual gets thirty-one and has only one sex chromosome. So it is male. Metaphorically speaking, the male grasshopper forms two sorts of sperms, male and female. The sex of an individual depends on which sort of sperm fertilises the egg.

On the whole, we should expect that when such a mechanism operates (Fig. 28), just as many males and females will turn up. However, there are circumstances which can influence the sex-ratio so that we meet with more or less divergent proportions and do not get

Race, Reason and Rubbish

exactly as many of one sex as of the other. Among human beings in general, 52 % of all births are boys[1] and 48 % are girls, but the figures vary somewhat in

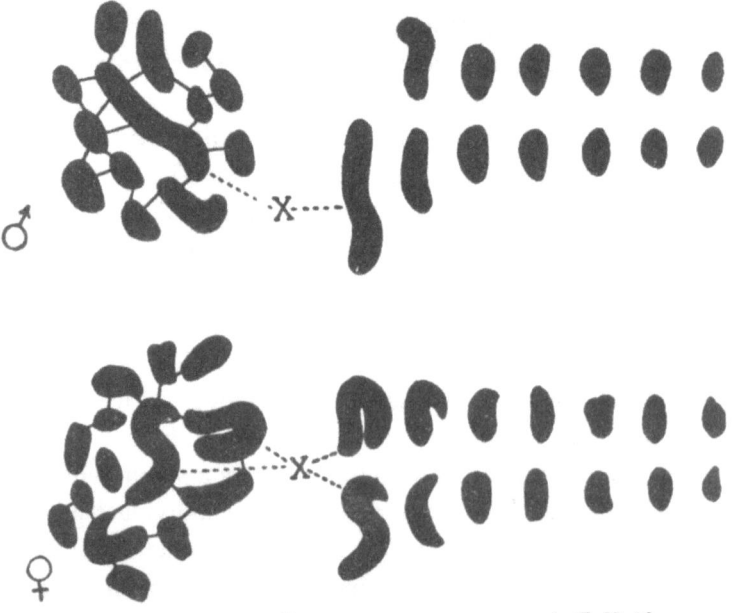

From THE THEORY OF THE GENE *by T. H. Morgan*

FIG. 28.—Chromosomes from the cells of the male (above) and female (below) of a plant bug (*Protenor*) showing the unpaired X-chromosome of the male

different circumstances. To mention only a few examples, the percentage of males is 49.3 among sheep, while it

[1] In vital statistics it is common to give the number of males per 100 females, i.e. 108, if 52 % births are male.

Race, Reason and Rubbish

is 52.3 among pigs and 51 among dogs. In part at least, this is explained by differences of viability among the two sorts of sperm or among embryos of either sex.

In principle, this is the way in which sex is determined in a great variety of animals; but there are several modifications of the basic plan. One such departure occurs in banana flies. In this insect both males and females have eight chromosomes (Fig. 17). Among males members of one pair, in contradistinction to all others of which both members are apparently identical, are visibly different. One of the chromosomes of this pair, called the Y-chromosome, is hook-like at one end. The corresponding member which has no hook is the real sex chromosome. Thus the banana fly produces two sorts of sperms, those with an X-chromosome and those with a Y-chromosome. If fertilisation occurs with a sperm containing the X-chromosome (or sex chromosome) we get an individual of the female sex, because it will have two sex chromosomes. If fertilisation involves a sperm having a Y-chromosome, we get a male, because the fertilised egg will have both the X-chromosome and the Y-chromosome. For a long while it was thought that the Y-chromosome contains no genes and is merely a dummy, there, as we might say, to keep up appearances. Subsequently it has been found out that particular genes can be located in it, including some which determine the power of fertilisation.

In cases so far surveyed the males produce two sorts

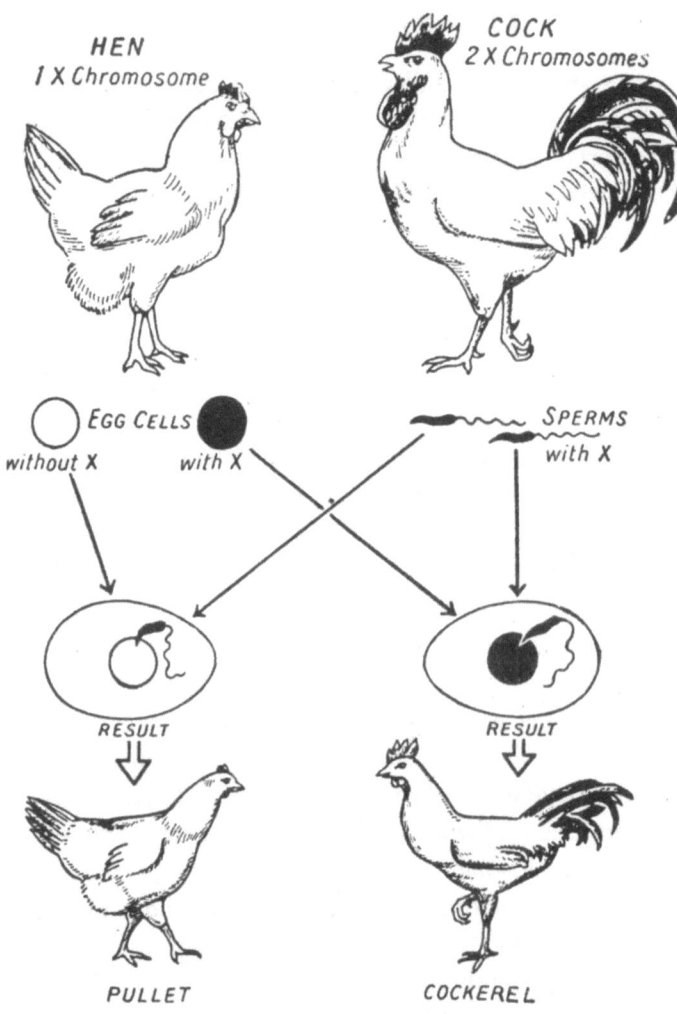

FIG. 29.—Sex determination in birds—the female is the heterogametic sex

Race, Reason and Rubbish

of gametes and the females only one type of egg. This state of affairs can be reversed. Among birds (Fig. 29), for example, the males have two X-chromosomes, while the females are *heterogametic*. That is to say, they have only one sex chromosome and therefore produce two sorts of gametes. This is also true of moths and butterflies. Among human beings, women have two sex chromosomes and men have only one. It thus happens that the sex chromosomes contain genes which influence the development of the individual. If two such genes are together, it gives development among human beings and many other organisms a bias towards the female condition. On the other hand, a single one gives development a bias towards maleness. If we look at the matter from the point of view of the gene itself the male is a low-grade female.

The intimate character of the mechanism is not the same in lower as in higher animals. Among some of the former we can say that every cell is definitely male or female. Among higher animals the connexion between cell composition and sex differences is more complicated. Genes which determine sex lead to the production of special glands which pour what are called *hormones*, i.e. special chemical products, into the blood. In their turn these substances influence the development of other organs in the direction of male or female differentiation.

Among lower animals, in which every cell has a distinctly male or female character, it may happen that

Race, Reason and Rubbish

a defective division occurs when a fertilised egg divides into several cells. Thus one of the products of cell division may get two sex chromosomes and the other newly formed cell may get only one of them. All the

FIG. 30.—Mosaic Wasp gynandromorphs. The dark regions have male characteristics, the light ones have the peculiarities of the corresponding structures in a female

descendants regularly subsequently produced from this pair of cells get one or two sex chromosomes respectively. So an individual is built up of two groups of cells, one male and the other female. The boundaries may also be defined in an irregular way. The animal can then be a mosaic of male and female cell groups. An animal of this sort is called a *gynandromorph* (Fig. 30). Among some insects of which the males and females have different colours, we can see from the colour which parts are male and which are female.

Among higher animals, the sex of which is determined by substances liberated from sex glands, we can interfere with sex development in various ways. By taking away the sex glands, and by grafting those of the opposite sex, we can modify differentiation, so that to all appearances an animal changes its sexual characteristics. If we only remove the sex gland, the result is that the individual often differs from both sexes. To some extent, though not consistently, it is midway between both of them. Eunuchs, i.e. men deprived of the sex glands early in life, have a type of body-build in many ways different from ordinary men or women. The earlier the removal of the sex glands, the more divergent are the eunuchoid characteristics. If a corresponding operation is carried out on a hen, it develops in the direction of maleness. It gets spurs, neck hackles and tail sickles, so that it is quite like a cock. This seems to imply that the body cells of the hen have a bias to develop in accordance with the male pattern when left to themselves. When the body is influenced by hormones of the ovary development is diverted towards femaleness (Fig. 31).

From VEREBUNGSLEHRE *by Ludwig Plate*

FIG. 31.—Castrated pullet with male spurs and sickle (above) and capon (castrated cockerel) with reduced comb but otherwise typically male

Race, Reason and Rubbish

Of late very significant discoveries have been made with reference to the more precise influence and chemical structure of the substances set free by the sex glands. These substances have an effect on the growth of the womb during pregnancy, on sex behaviour, and on the periodic processes which are connected with the maturation of the egg and milk production after parturition. We also know that they influence body-build during the course of development. The circumstance that twins produced from different eggs may be of opposite sex throws light on how these hormones work. Cows sometimes bear twins of which one is a bull-calf and the other is halfway between a male and female. The latter are called *"free martins."* For a long time people have discussed which sex such a calf belongs to, i.e. whether it is a defective bull-calf or a defective female. As far as its inherited make-up is concerned we now know that it is the latter.

When twins produced from two eggs turn up among cows, it often happens that there is a short-circuit between the blood circulation of their placentae, so that the two foetuses exchange blood (Fig. 32). If one of them is male a comparatively vigorous liberation of hormones begins early. The one which is female produces hormones which are less effective. The male hormone wins the day, and the result is that the female foetus is biased towards maleness and becomes a half-way house, i.e. a free martin.

Race, Reason and Rubbish

Besides genes which determine sex, sex chromosomes contain others. When these genes accompany the X-chromosomes, they must in some way depend on the sex of the individual to which they are transmitted. In May 1910, Morgan found among banana flies a

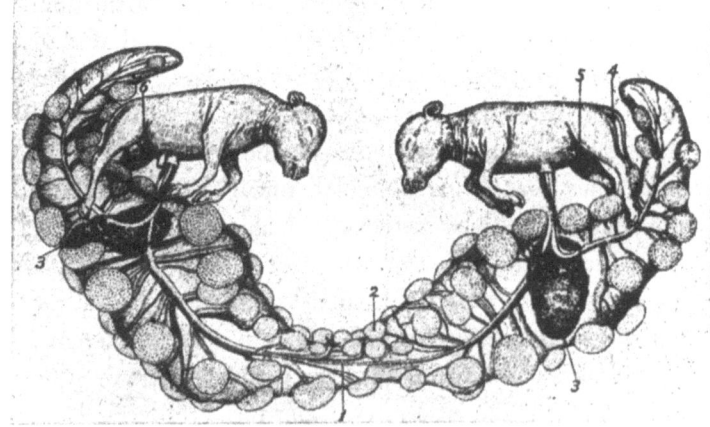

From THE BIOLOGY OF TWINS *by H. H. Newman*

FIG. 32.—Common foetal circulation of cattle twins

solitary male with white eyes instead of red ones, which are normal in this species. He mated this male with an ordinary red-eyed female and got an unexpected result from repeated crossing. In the first generation the offspring were entirely red-eyed. The gene for red is therefore dominant over the gene for white eye colour. When the cross-bred flies were mated among themselves, only red-eyed females were obtained, but half the males had white and half of them red eyes. As we expect in

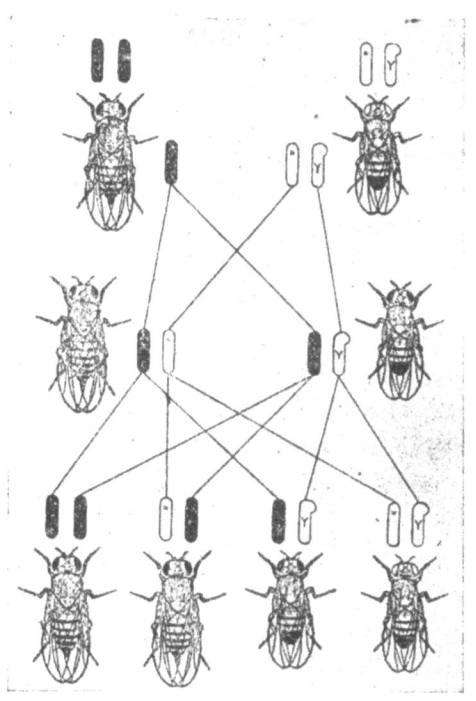

From THE THEORY OF THE GENE *by T. H. Morgan*

FIG. 33.—Crossing-over in transmission of two recessive sex-linked characters of the Banana fly—mating of pure wild type female with red eyes and gray body to double recessive male with white eyes and yellow body

dominant inheritance, one quarter of the progeny were white-eyed, that is to say one quarter showed the recessive character. What was peculiar about it was that all the flies with white eyes were male, and that Morgan never got a white-eyed female in such crosses.

If we concede that the gene for white eye is on the X-chromosome of the white-eyed male first discovered, it is easy to understand the way in which inheritance occurs (see Fig. 33). In the first hybrid generation, the paternal sex chromosome carrying the mutant gene can never exist by itself in an egg. All fertilised eggs of the second generation must contain a chromosome with the gene for red eye derived from the female of the parental generation. So the cross-breds are made up partly of males which have a gene for red eye and are red-eyed, and partly of females with two sex chromosomes, of which one contains the gene for white eye and the other the gene for red. All the flies are red-eyed because the red-eye gene is dominant.

A sex chromosome with the gene for white eye can be present by itself in the fertilised egg of the next generation. Females of the second produce two sorts of eggs—partly eggs with a sex chromosome containing the gene for red eye, partly eggs with a sex chromosome with the gene for white eye. Males produce two sorts of sperm. Some have a Y-chromosome and others an X-chromosome with the gene for red eye. If fertilisation involves the latter type of sperm, an individual develops with two chromosomes of which at least one must have the gene for red eye. Half of the fertilised eggs have another sex chromosome with the gene for white eye, but the latter can never make itself felt, if accom-

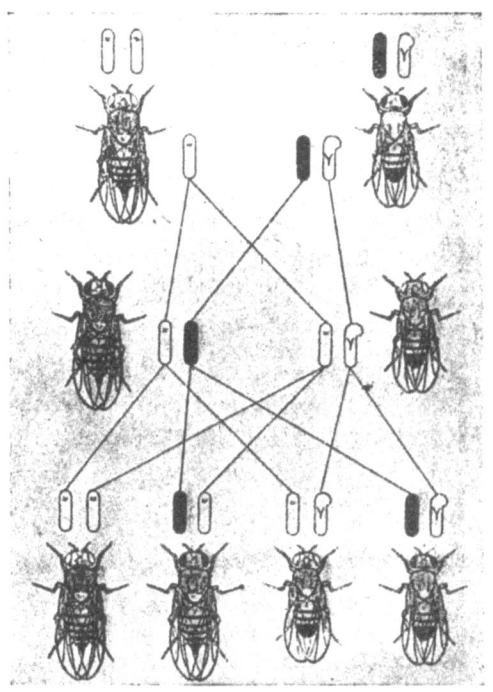

From THE THEORY OF THE GENE *by T. H. Morgan*

FIG. 34.—Crossing-over in transmission of two recessive sex-linked characters of the Banana fly—mating of yellow male with wild type (red) eyes to white-eyed female with wild type (gray) body colour

panied by a second one carrying the red-eye gene. All the females are therefore red-eyed, and half of them are homozygous. If fertilisation is brought about by a sperm which contains the Y-chromosome, the fertilised egg will only contain one sex chromosome. In half of them it will carry the gene for white eye. So half of the males will be white-eyed, and if mating is carried out in this way, white eye colour turns up in males only.

We can also arrange crosses in such a way as to get white-eyed females (see Fig. 34). If we cross a white-eyed male with one of the red-eyed females which have a single gene for red eye colour, we get females of which half are white-eyed and males of which half are also white-eyed. The heterozygous females make two sorts of egg—partly eggs which contain a sex chromosome with the white-eye gene and partly eggs in which this chromosome has the gene for white eye colour. The male makes gametes with a Y-chromosome and others with an X-chromosome having the white-eye gene. It is easy to figure out what results will happen when the gametes get together.

The transmission of characteristics which depend on genes in the sex chromosome is said to be *sex-linked*. The peculiarity of this sort of transmission is that when—as is usual—the male is the *heterogametic* sex the presence of a recessive sex-linked gene in the male is always manifest. When such a male mates with a normal female it is suppressed. It exists among the females only, and in them it is latent. From such females it can pass over to males in the next generation. When, as in fowls and moths, the female is the sex with only

Race, Reason and Rubbish

one chromosome, the picture is reversed. Such crosses, involving sex-linked characteristics, have been used as a convenient device for recognising sex at an early stage. If we pair *black* Minorca cocks with *barred* Plymouth Rock hens, we always get cockerels with white stripes on the neck, and these can be seen immediately after hatching. So criss-cross inheritance can be put to a practical use.

For a long time we have known about some human characteristics which are inherited in this way. The best-known example is colour-blindness. If we go into the family history of a colour-blind man we find that a male ancestor (generally the grandfather) was colour-blind. The trait is transmitted from the grandfather through the daughters to his grandsons. Colour-blind women do exist, but they are comparatively rare. A man exhibits the character if he gets the gene in his single sex chromosome. If she is to be colour-blind, a woman must get it in both of her own. Just as the likelihood that we win a lottery prize once is much greater than the probability that we do so twice in succession, the probability that an individual will get two genes is less than the probability that he or she will get only one of them. So women who are colour-blind turn up only in relatively few cases, that is to say when a woman who is the daughter of a man who is colour-blind, and therefore has the gene for colour-blindness, herself marries another colour-blind man. What then

Race, Reason and Rubbish

happens is like the situation already explained in what has been said about the banana fly.

In this connexion it is necessary to distinguish between inheritance which is *sex-linked* and inheritance which is *sex-limited*. In sex-linked inheritance we are concerned with genes which lie in the sex chromosomes. In sex-limited inheritance we are concerned with characters which only appear in one or the other sex. The genes need not lie in the sex chromosomes but generally in the other chromosomes collectively called *autosomes*. So they may be transmitted in the ordinary way. If a gene determines such a characteristic as milk-yield, it can never display itself in males. We can get a first-hand estimate of the milk performance of a cow and hence a clue for finding its gene equipment. The equipment of a bull in respect to the genes which are responsible for milking capacity can be judged only by its ancestry, e.g. its mother's and its paternal grandmother's performance, or by gradually collecting reports of the performance of its offspring. This is one of the difficulties of stock improvement.

We have learned that Mendel made the fundamental assumption that the gene is a fixed entity which may be present or absent in a cell but cannot be there in greater or less degree. This conclusion still stands, but researches carried out by Richard Goldschmidt have shown that different genes with the same effect can have different intensities. At first sight the consequences of this may

From MECHANISMUS UND PHYSIOLOGIE DER GESCHLECHTSBESTIMMUNG *by R. Goldschmidt*

FIG. 35.—Sex intergrades of the Chocolate moth showing various gradations between the pale female (*a*) with thick abdomen and filamentous antennae to the dark male (*k*) with thin abdomen and feathered antennae

Race, Reason and Rubbish

seem to be inconsistent with our previous assertion. The situation is as follows. Goldschmidt has made a study of crossing between different species of the chocolate moth which is widely distributed throughout the world, but has more or less divergent local forms. Among other things he has mated Japanese with European forms. It turns out that if Japanese males are crossed with European females normal males are produced, but the female offspring develop in a way intermediate between males and females (Fig. 35). They are therefore called *intersexes*. The intersexual condition is illustrated, for instance, by the colour pattern. Normal females are white, while males are gray-brown. The intersexual females are gray-brown, but with more or less extensive white flecks. By crossing different varieties Goldschmidt got different degrees of transformation from male to female. He succeeded in getting good evidence for the view that genes which determine sex, act with different intensities in different varieties of the moth.

When a Japanese male is mated to a European female, the offspring which will be female get their *single* (see p. 95) sex chromosome from the Japanese father. The X-chromosome does not contain genes strong enough to compel the individual to develop in the female direction. The quantity of sex-determining material in the Japanese male is not strong enough to compensate the influence of the autosomes derived from the Euro-

Race, Reason and Rubbish

pean mother. At first development goes in the female direction (Fig. 36), but afterwards it takes a turn towards maleness. So long as we cross individuals within one and the same local variety the quantities of sex-determining genes are appropriate for the production of

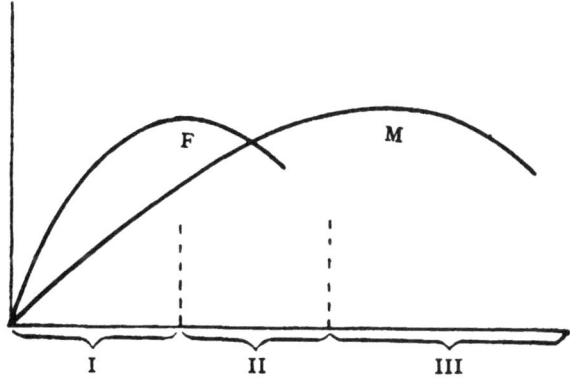

Fig. 36.—Diagram to represent Goldschmidt's interpretation of sex differentiation. The vertical distances stand for the amount of male (M) and female (F) sex determining substance; the horizontal distances for time development. In a normal male and a normal female M or F substance respectively predominates during Phase II when the developing organs take on their sexual characteristics. In a female intersex (here shown) the male determining substances get the upper hand before Phase II ends

normal males and females, but they are not suitable when crosses are made between different varieties. None the less, Mendel's assumption is right. A single gene cannot change its size. It is present or lacking but it does not increase or diminish in the process of trans-

Race, Reason and Rubbish

mission; but genes in different individuals, possibly in different pedigrees and in any case in different subspecies, can be more or less numerous, and can therefore produce effects of different intensity.

One detail of special significance is this. Among intersexual females development at first goes in the male and afterwards in the female direction. So we have reason to imagine that the influence of one group of genes rises quickly and afterwards subsides, while that of the other rises more slowly to its maximum (Fig. 36). In such circumstances the rate at which individuals grow must play a part in deciding their sex characteristics. If the individual develops quickly it should be full-grown while the influence of the female-determining genes prevails. This will confer on it the characteristics of the female sex. If growth is slower, development should first go in the female and then strike out in the male direction. It has been possible to influence the sex of a worm called *Bonellia* by an abnormally rich diet. If we give it rich food the animal grows quickly and becomes a female. If it gets a scanty diet it does not reach full stature during the time when the female-determining genes are effective. The development first proceeds in the female, then in the male direction, and the more the animal is starved, the more decidedly does it take the latter.

In nature the bee found this trick long before we did. What happens among bees is that males develop

Race, Reason and Rubbish

from unfertilised and females from fertilised eggs. If a larva produced from the latter is starved it develops along the road to femaleness to a certain point, but it does not complete its growth before development is deflected towards maleness. The result is that the individual which results from the treatment is an intersex, a half-way house between male and female. Such a bee is called a *worker*. Worker bees are free from sexual preoccupations and can therefore devote themselves more fully to the monotonous tasks of communal life. Bees arrange communal feeding so that the majority of individuals are semi-starved intersexual workers. In any single bee-hive there is only one well-fed female or *Queen-bee* who is the focus of communal existence. By degrees the stock of fertilised eggs gives out. The queen lays a smaller percentage of fertilised eggs and males are born from unfertilised ones. A new female from one of the fertilised eggs is then fattened up. It swarms with the males, and after it has been fertilised becomes the queen bee of a new community. The males are destined to die after the marriage flight, and the new swarm consists of the queen and a large number of intersexual individuals, the under-fed workers, which live exclusively for the society of which they happen to be members.

Perhaps we might say that in the bee-hive the ideal society of some eugenists has been realised. Reproduction and sex determination are completely regulated. Sexual connexion is only allowed when the good of the community demands it. Individuals live for the State and are nurtured to produce the character which the community is considered to need.

CHAPTER V

IF simple Mendelian inheritance occurs, a gene produces, singly or in duplicate, a particular character, and therefore has a characteristic effect. This is one of Mendel's basic principles. Like other scientific generalisations, if correct, it is correct only in particular circumstances. On its own account a gene can never give rise to an individual. To produce an individual a whole group of genes is necessary. Besides this, there must be appropriate external conditions to implement development. Even if a gene decisively determines the development of a character in most situations, external conditions may influence development one way or the other in special circumstances. Moreover, the other genes in the hereditary make-up of the individual play a significant role. Goldschmidt's research, which we have talked about earlier, shows that, while some genes determine sex in normal circumstances so that we get normal males, the same ones may not have a strong enough effect, if they get into a new gene complex, and so yield intersexes. This implies that the influence of a particular gene depends on that of another.

Much the same applies to characters other than sex. Presumably genes which determine mental disease also depend to some extent on other genes. In particular pedigrees we sometimes find an individual who is too

Race, Reason and Rubbish

original, has queer ideas, a cut-and-dried bearing, and so forth. In the same family there may be another one who is clearly insane, has fixed ideas, and such a peculiar way of carrying on that he has to be taken in hand and committed to an asylum. Conceivably, both of them have the same gene (or genes) for insanity, but one may have got genes which determine better intelligence. With their help he can get on top of his abnormal tendencies. To be sure, he may be a bit eccentric and odd, but his departure from normal standards may not be so great that we have to certify him.

In general, we may therefore say that Mendel's own statement of his theory involves a simplification of what really happens. When we are concerned with how a gene works we ought really to take stock of the whole gene complex in which any particular gene is present. Genes co-operate in a very complicated way. An important detail which shows that this is so, is the quite recent discovery that the effect of a particular gene depends on its position in a chromosome. In special circumstances it may get shifted, so that it does not lie in its normal situation, but is surrounded by different genes. Its effect may then be altered. Such *translocation* can take place when chromosomes are damaged, as, for instance, by X-rays. When this happens, we say that the gene has a particular *position effect*, meaning that it produces one effect in one position and another effect in a different one.

Race, Reason and Rubbish

This emphasises that genes are not really such independent entities as we used to think. The chromosome is individual and the genes are chemical constituents which react with one another so that the result depends on the whole. Individually the separate parts of a chromosome can do nothing at all. We might compare a chromosome to a single chemical molecule of which the genes correspond to groups of atoms. This way of looking at the question is linked up with known phenomena of organic chemistry, but, to say the least, it is doubtful if it is more than a metaphor. To get a better grasp of the nature of the genes some people have tried to connect them with substances called *catalysts*, which influence chemical reactions without being used up in the process. Although we cannot ordinarily ignite a lump of sugar, there is no difficulty in getting one to burn if we throw a little cigarette ash on it. The ash acts as a catalyst. If the ash is there, ignition takes place when a moderate temperature is reached; and the ash produces its effect by its presence alone. There are organic substances, called *enzymes*, which work in this way. Their presence ensures that chemical reactions which could not otherwise occur, or, more precisely, reactions which go on so slowly that they have no significance, will take place in the body.

Of course, we must remember that this is a matter of opinion. It helps us to picture the processes which we are discussing, but the comparison must not be

Race, Reason and Rubbish

taken literally. There may be decisive differences between the basic processes of living tissues and chemical reactions which occur among dead matter. What separates living from dead matter remains obscure. Eventually, we may hope to get fundamental enlightenment about the intimate nature of living processes through investigations about the way in which genes do their work. Just as atomic research leads on to questions about the nature of physical matter, research on heredity will gradually come up against questions which border on the nature of life itself.

Meanwhile it goes without saying that the questions here dealt with are still far from clear, though there are some positive results to which we can already point. We can specify genes which work mainly in two different ways. One type of gene influences cell division. Such genes decide whether an organ will become large or small. If a gene has a powerful influence on cell division the organ must be comparatively large. Maybe such a gene can affect only parts of an organ. All genes which influence size and form may therefore be said to speed up or slow down cell division. A second type of gene influences the nature of the cell body. A gene which determines human eye colour produces its effect by building up pigment in the cell body around the nucleus. We see the black membrane which clothes the posterior aspect of the pupil through a layer of semi-translucent cells. If much pigment is formed in the pupil, the eye will be brown. If pigment is lacking it will be blue or gray. When the result is blue-gray, it may be so for the same reason that distant forests look bluish when we see them through a thicker layer of half transparent air. In such circumstances we ought to speak of *qualitative* genes which affect the character of the cell in contradistinction to *quantitative* genes which affect the number of cells in the organism as a whole or in different parts of it. Meanwhile it is still a mystery how genes can really act so as to confer a different character on different parts of an organism.

We have learned that the chromosomes split lengthwise in ordinary cell division, and how this ensures that both newly-formed cells which result from it get exactly the same equipment of genes. Only the reduction divisions are an exception to this rule. Now if all cells get exactly the same equipment of genes, we might expect all cells to be exactly alike. The organisms which develop from a fertilised egg would then consist of a greater or smaller collection of cells which would not be visibly distinguishable. In real life some cells develop as cells of the liver, others become muscle fibres, nerve cells, and so forth. We are therefore up against a paradox. The solution must be that even if the chromosomes divide in a completely symmetrical way, the same cannot be said about other parts of the cell.

Before the fertilised egg begins to develop there must be some sort of localisation (Fig. 37), so that for instance those parts which lie at one end of the cell have a chemical composition different from those which lie at the other. Cell division may occur in such a way that, when the chromosomes divide, one half are drawn to the former and one half to the latter. Each set then builds a new nucleus, and two new cells will be produced with *cell bodies* built up of different constituents. So the identical gene complex of the two cells has different

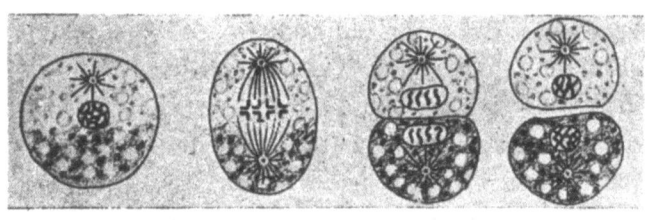

FIG. 37.—Diagram showing how unequal distribution of extra-nuclear cell contents leads to the production of cells in which *identical* chromosomes react with a different substratum of other material

building materials at its disposal. Through repeated division of this kind, differences are intensified, and the same gene complex of one type of cell will work on a chemical substratum quite different from that of another. In this way we can explain how we have different types of cells in our bodies in spite of the symmetrical distribution of the genes themselves.

In different species of animals the asymmetrical stratification of the cell body may begin earlier or later. In some species it already exists at the time of fertilisation.

Race, Reason and Rubbish

This can be shown by cutting away a small piece of the fertilised egg. The result is that the individual which develops will lack some region of the body or a particular organ. Among other species nothing happens if part of the egg is cut off. The individual which develops is not defective. Thus we can make a distinction between the *mosaic* egg and the *regulation* egg. Mosaic eggs may be said to have the appearance of a mosaic of the various parts of the adult itself. The regulation egg, on the contrary, has not reached such an organisation. The cell body is homogeneous, and if we remove part of it, it regulates its own development so that no deficiency results.

The effect which a particular gene brings about does not merely depend on the rest of the gene equipment or on the chemical properties of the cell body. External circumstances in which the individual develops can also play a greater or smaller role. It cannot be too strongly emphasised that we do not inherit the character itself. What we do inherit is merely a gene which determines its presence, and the situation is not so simple as it would be if a gene always produced a particular character. A gene can produce a particular character in a particular environment and another character in a different one. A good example which illustrates this is the following. There are so-called Russian rabbits (Fig. 38) which have black paws, black eyes, and a black snout, but are otherwise white. A German geneticist hit on the idea

Race, Reason and Rubbish

that the reason for the occurrence of the black pigment might be that the more remote ones have a lower temperature than other parts of the body. To test

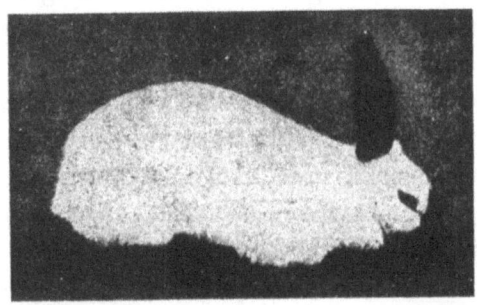

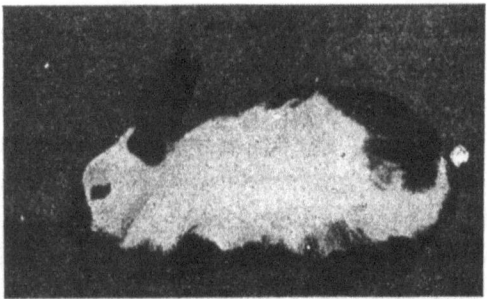

From MENSCHLICHE ERBLICHKEITSLEHRE *by Baur-Lenz-Fischer*

FIG. 38.—Normal Russian rabbit (above) with black pigment confined to the extremities, and individual of the same variety kept in cold after shaving the regions where the new hair is now black

whether this was so, he shaved the hair off the back of a rabbit in mid-winter and let it live in the open air. The hair which grew gradually on the shaven region

Race, Reason and Rubbish

exposed to the cold became black. In other words, the rabbit had genes which determine that black hair will develop on parts of the body at a low temperature, and white hair in regions at a higher one. By bringing up Russian rabbits in warm surroundings, we should presumably be able to get them quite white, but so far as the author knows no such experiment has been carried out.

It is really a very important practical problem to define within what limits individuals with a particular gene equipment can vary under the influence of different external circumstances. One method of carrying out such analysis depends on the use of so-called pure lines. The best thing for this is an organism which is self-fertilising, for instance homozygous brown beans. Research with the latter has been carried out in the following way (Fig. 39). Among those beans which come from a single plant the largest and the smallest were chosen. They were cultivated; and the characteristics of the beans obtained have been investigated. Selection has been continued from one generation to another, and has no effect. The beans obtained vary with respect to size, but the mean size is the same in successive generations in spite of the fact that we may persistently choose particularly large or particularly small ones for further breeding. Researches of this sort show that selection from genetically homogeneous individuals is ineffective. The environment can influence the plants so that the

Race, Reason and Rubbish

beans are smaller or larger, but this leaves no permanent trace and does not affect the gene complex. Thus it has been possible to show that the latter remains fixed, as also to get a rough idea of the extent of variation which

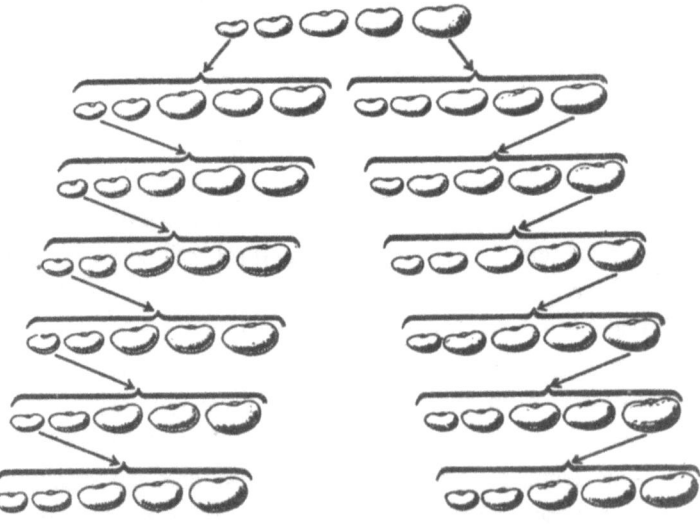

[FIG. 39.—Diagram to illustrate Johannsen's experiments on beans (p. 125), showing that selection within pure lines has no effect on the character of the offspring

ordinarily depends on the fact that external circumstances operate in various ways.

To disentangle what significance the genes have and what range of variation can be brought about by external circumstances is specially important in connexion with human beings. We all agree that genes

Race, Reason and Rubbish

play a part in producing differences which distinguish different individuals, as we also agree that external conditions, such as diet, education, and the like, have an essential influence; but truisms of this sort are of scant value. We need a more exact measure of how much the gene does in comparison with the milieu. It is, of course, extremely difficult to arrive at more definite conclusions, but one line of attack which is used more and more comes from researches upon identical or uniovular twins. Unlike diovular twins which are produced from separate eggs, such twins are produced from a single egg which gives rise to two separate individuals, instead of to two halves of the same body. They have therefore the same hereditary make-up, and—with certain exceptions that cannot be dealt with here—differences which distinguish them must be attributed to the influence of the environment. One twin, for instance, may get better food, and is therefore stronger and taller. The other gets somewhat inferior food, and is therefore not so well developed.

Differences to which uniovular twins are exposed are not usually great. During the period of growth they customarily get the same food, the same early training, or school teaching; and in most respects uniovular twins are amazingly alike. In appearance, as is well known (Fig. 40), they are so much alike that they are often mixed up. Usually the mother can tell them apart, but the father may mistake them. Schoolmates and

Race, Reason and Rubbish

teachers often confuse them. Sometimes there is a small difference which can give a clue. One can have a clockwise hair-whorl, the other anti-clockwise. Investigations

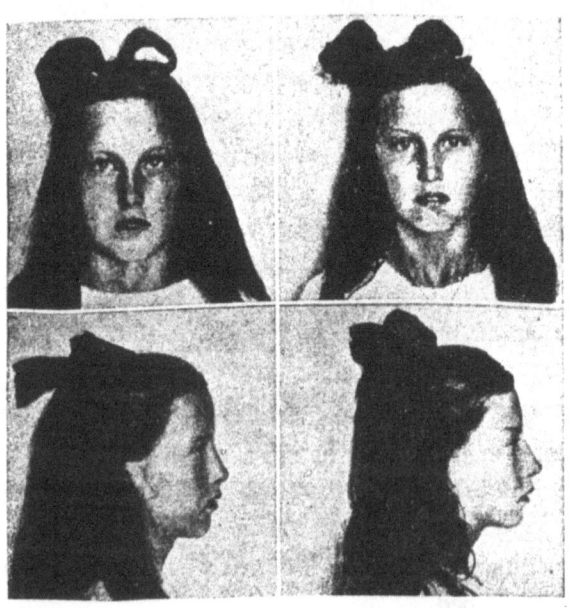

From TWIN BIRTHS AND TWINS FROM A HEREDITARY POINT OF VIEW
by Gunnar Dahlberg

FIG. 40.—Uniovular (monozygotic) twins

into differences with respect to stature, arm length, head form, breadth and length of face, etc., have proved that even if they are very much alike in appearance, there are still essential differences between them. With the help of these differences between uniovular twins

Race, Reason and Rubbish

and corresponding differences among diovular co-twins whose hereditary make-up varies to the same extent as that of ordinary brothers and sisters (Fig. 41), it has

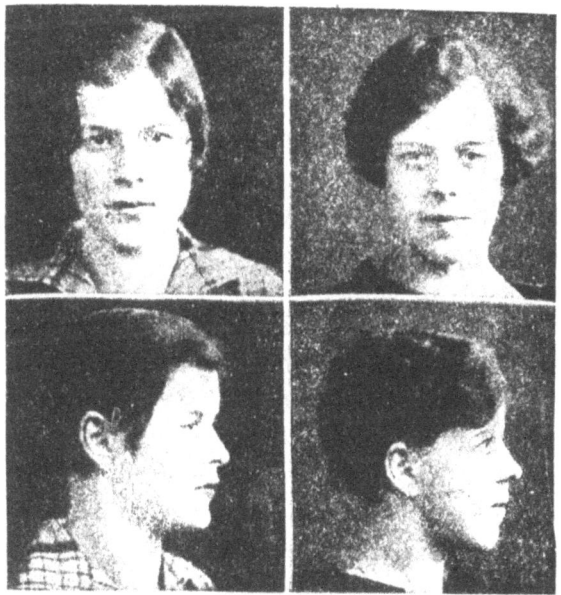

From TWIN BIRTHS AND TWINS FROM A HEREDITARY POINT OF VIEW
by Gunnar Dahlberg

FIG. 41.—Diovular (dizygotic) twins

been possible to make a rough estimate of the extent to which body-build is determined by nature and nurture. The result is that differences of genetic equipment account for roughly twice as much as differences of nurture. This means that if all people in a population had the same genes, and if the general, social, and physical environment did not vary more than the environment to which the two members of a pair of uniovular twins are exposed, there would be a reduction of the spread of variation with respect to body-build. The latter would be half as great as the spread of variation which would exist if everybody in a population with gene differences as great as those which generally distinguish diovular twin pairs were brought up in exactly the same environment.

In this connexion we ought to remember that the gene equipment of diovular twin pairs does not vary so much as in the population at large. Twins produced by two eggs have the same parents, and though this does not prevent them from getting different genes, the variability with respect to gametic constitution is not so great as may exist among other individuals who have not the same parents. So we should not therefore over-value the results which can be got from research on twins. This does not alter the fact that investigations dealing with twins have a very great interest, and not the least when the inheritance of disease is involved. Where diseases are concerned uniovular twins exhibit a surprisingly close agreement, while differences among diovular twins are very considerable. The same applies to mental characteristics. By the use of intellectual ratings it has been possible to show that differences of mental characteristics, measurable by tests more or less appropriately called *intelligence tests*, are comparatively small among uniovular but significantly greater among diovular twins.

In the opinion of the writer, one observation which comes out of such work is that temperamental differences are comparatively great among uniovular co-twins. Parents, teachers, and other persons in close relation to them, often state that one is cheerful and obedient while the other is reserved and defiant. In so far as this is so, it means that temperament is affected to a relatively large extent by extrinsic circumstances. However, it is not yet possible to make exact measurements of those characteristics which we signify when we talk of temperament. This opinion is mentioned merely to seize the opportunity of recalling a special difficulty which arises in research on human inheritance. Among other characteristics which interest us most from the standpoint of heredity, we must include mental ones; and research on the latter is tremendously cramped by our lack of precise methods for measuring them. So no results of scientific value have yet been reached in this connexion.

CHAPTER VI

INTEREST in genes which distinguish people and human populations began rather late. It was first focused particularly upon the fact that nations and empires which have reached a high level of culture decline and go under. People asked themselves the reason for such degeneration, and historians spoke of inherent decadence, moral looseness, the spread of pleasure-seeking, and the like. The phenomenon must have some explanation, and it was taken for granted that populations had gone downhill—or, more loosely, that *races* had done so. So the problem was to find out the cause of racial decline. This brought into being a new type of enquiry called race biology. The name struck root—most of all in Germany.

In England interest in the problems of human heredity had begun at an earlier date. Among other questions one had specially attracted attention. Some people were curious about how individuals who have got into influential positions are related to one another, and so forth. Obviously it is difficult to decide what part inherited endowment plays in this, because advantages which depend on unequal distribution of property play an influential role in contemporary society. Circumstances of this sort have a special significance in England, where the extraordinary organisation of school life

Race, Reason and Rubbish

excludes children of the less wealthy from prominent social positions. Naturally, questions were also asked about characteristics which have no direct connexion with differences between social classes. Such interest, focused on persons of ability, sticks out in the very name *eugenics*[1] which the new enquiries earned in England.

The foundations of scientific work on these lines were laid down by Darwin's cousin, Francis Galton. Like Mendel, the latter was interested in mathematics without being equipped with more intimate mathematical knowledge; but he secured accomplished co-workers, foremost among them Karl Pearson. The trend of investigation which flourished in England is referred to as biometry. It proceeded from the assumption that there is some connexion between the characteristics of parents and children. We now know that this connexion may depend on similarities of the environment, but can also depend on biological inheritance. The *biometrical* school tried to disentangle the nature of the connexion by statistical calculations (*correlation coefficients*), and the work assumed a mathematical character. Few results of practical interest emerged; but it produced statistical methods of greater refinement and practical usefulness in research.

Meanwhile, a school of different aspect was built up on the continent, and especially in Germany. People

[1] I.e. study of the well-born.

Race, Reason and Rubbish

began to apply to human beings results already obtained by Mendelian methods in zoology and botany. Since we cannot carry out experimental crosses between human beings, we can only try to imitate experimental matings by searching through pedigrees. Material that could be got was often defective and, above all, of limited range, because human families are comparatively small and we meet with great difficulties in getting information about ancestors who lived several generations back. Meanwhile some interest in race classification and race problems survived from a pre-Mendelian generation. Through investigations influenced by the experimental outlook results of real significance were reached. In particular it was successfully shown that Mendel's laws are applicable to human beings.

Opposition between the biometricians and the continental Mendelian school persisted for a long while. The former treated the continental scientist with some contempt, because his methods were defective from a statistical standpoint, and because of his disposition to draw conclusions from insufficient material. The Mendelians considered that the English workers were not in step with time, because they did not apply the quantitative results of Mendel's teaching. They pushed prudence too far when they tried to solve problems by mathematical rule of thumb. They refused to make more assumptions than they regarded as absolutely essential, insisting that Mendelism was an unproved

Race, Reason and Rubbish

hypothesis. Of late, research of this sort has taken a new turn through a union of both methods. There is a growing tendency to carry out investigations in which mathematical and statistical methods are used to unravel consequences which Mendel's laws imply when applied to human populations in the near or distant future.

By keeping experimental animals or plants in precisely similar surroundings, it is generally possible to overlook the influence of environment in research on inheritance among plants and animals. In research on human heredity we have to take stock of the significant role of external circumstances which we cannot alter. The influence of heredity itself can vary within wide limits, but if divergence from the norm is too great an individual cannot develop. Genes which produce abnormalities that are too great may lead to an individual malformed in one way or another so that it is not viable. So also, external circumstances may differ within wide limits. Diet, temperature, fresh air, and light, are all of them agencies which can vary to some extent. If the milieu departs too far from the norm, as it could if the earth itself were to become too hot or too cold, no human beings can develop. Once more, mortality sets a boundary to environmental variation, but inside this boundary nature and nurture react in many different ways. Clearly, it is very difficult to analyse the tangled skein of agencies opposing and reinforcing one another.

Race, Reason and Rubbish

An appropriate classification of all situations is as follows.

1. *Genetic characters*, in the strictest sense of the term, are those that depend on genes and do not appreciably vary under the influence of external circumstances. If, for example, a man has the gene for colour-blindness he will be colour-blind, however he may be nurtured and whatever medicine or food he gets. Although there is always a possibility that unknown agencies might act on the development of the individual to prevent the appearance of the character, normal ones have no effect.

2. *Environmental characters*. To this group belong characters which depend on particular external circumstances. Whether a man contracts typhus depends on whether he has been infected, but if a dog is infected with the typhus organism it does not sicken. This indicates that human beings have a liability to typhus which is genetically determined, and that dogs lack it; but since the genes concerned are present in nearly every one, it is proper to overlook the genes themselves and to emphasise the significance of the environment alone when we wish to be brief. The boundary of such characters is not always a sharp one. Isolated individuals can accommodate the typhus organism without sickening, and are therefore carriers of the bacillus.

3. *Conditional characters*. This group includes characters that appear only if there is a combination involving

Race, Reason and Rubbish

a particular milieu and particular genes. The majority of people who are exposed to infection by polymyelitis, a form of infantile paralysis, do not get ill. Presumably, only 2 or 3 % among younger ones are liable to get it. Thus the sickness turns up when individuals with certain genes which are not widespread are exposed to special external situations.

Naturally, there is no very sharp boundary between these three groups. Whether inherited disposition or infection is more or less important, is a matter for discussion, where some illnesses are concerned; but even if the outlines of our classification are blurred, it has still some value from a practical point of view. The aim of research into inheritance is mainly to clarify the role of the genes themselves. So it is most important for us to try to get a clear perspective about genetic characters. It is more difficult to deal with the role of heredity in our third group of characters. So far as the second is concerned, the issue is clear because the appropriate genes are generally present.

It is not surprising that initial attempts to disentangle the role of genes in human affairs were directed to diseases. The latter are comparatively easy to investigate—partly because they are so decidedly different from the normal condition. The characteristics of disease are momentous for those who come up against them, and from this point of view they are very important. From the standpoint of the population as a whole,

Race, Reason and Rubbish

genes which determine intelligence play an even more significant part in everyday life. At a low level of culture mere brawn was a more important characteristic than it now is. For civilised folk the general intellectual level and the frequency of specially gifted individuals is more important than average muscular strength or the more exceptional accomplishments which distinguish sportsmen of the international class.

With regard to disease, scientific work was first directed to the outside world. This was right and proper, because we can take preventive measures against external agencies, and thus get results of practical usefulness. We have now got so far with this that even hereditary diseases are beginning to excite interest. As we are more successful in preventing diseases which depend on external circumstances, the class of diseases wholly or mainly determined by genes must become increasingly important. When we get as far as imagination can reach, we may be able to prevent all accidents, all infectious diseases, all sicknesses which depend on unsuitable diet, and all mental disturbances which depend on unsuitable upbringing. Perhaps when this happens we shall be left with nothing but diseases which depend on inherited disposition. Little by little the problem of how to prevent genes for hereditable diseases and defective resistance from passing over to the next generation will therefore assume much greater importance.

Race, Reason and Rubbish

Clearly the problem of how genes for mental characteristics which constitute a good endowment, that is to say, talents of different sorts, capacity for work, and so forth, are also of fundamental importance. While a community remains at a primitive stage, people of rather low intelligence can acquit themselves comparatively well. As our social life becomes complicated, life becomes more difficult for those who are afflicted with lowly intelligence or a less well-balanced temperament. In this sense we may even go so far as to say that ill-adjusted individuals are perhaps more common in higher types of civilisation. If so, they are going to be not merely a serious problem but a real danger to society.

So far, we have tacitly adopted a social point of view in relation to the problem of human heredity, but we can really approach it in two ways. First, we can look at it from the individualistic standpoint. Confronted with a person who has particular traits, and knowing also that among his or her ancestors or other relatives there are individuals with particular characteristics of one kind or another, we can ask ourselves the following question: if such a person marries, what is the risk that his or her offspring will get such characteristics, as, for example, particular hereditable diseases. In so far as we know more about the pedigrees of both parties to the contract, and more about the mechanism of inheritance which determines the relevant characteristics, we shall be in a better position to give a prognosis. We can

Race, Reason and Rubbish

also look at the problem from a collective standpoint, and ask what is the risk that the gene-equipment of a population will be altered in one direction or another in particular circumstances.

In this context we shall not deal with the first way of putting the problem. The different situations which are imaginable are extremely variable. We should have to discuss a large number of characteristics and diseases of which the mode of transmission is known. Even a superficial treatment of such matters demands much space, and those who want enlightenment would do best to consult an expert. The social statement of the problem can naturally be split into many subsidiary issues. So far, scientific work which has been carried out in connexion with such questions is still in its infancy. Many of them are unsolved. In what follows we shall try to get a brief grasp of the elements of what is already known in this field.

When we discuss populations in relation to inheritance we have to take some things for granted as a basis for further argument. We may start with the assumption that mating occurs at random and that individuals of different types reproduce their kind to the same extent. In such circumstances all genes have the same chance of being transmitted to another generation. The genes themselves combine at random, and a population in which mating and reproduction go on in this way must stick to a fixed make-up from the standpoint of heredity.

Race, Reason and Rubbish

No changes happen. Individuals of different types turn up in the same proportion in successive generations. Scandinavian geneticists speak of this state of affairs as

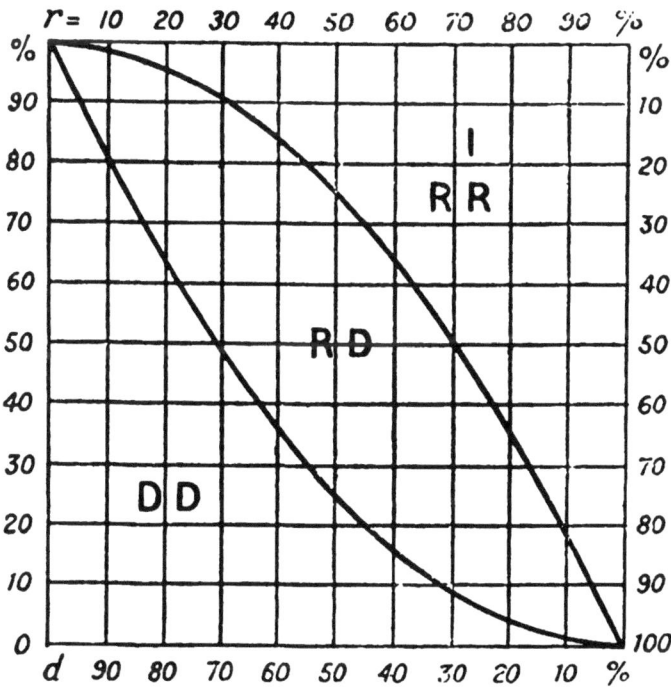

FIG. 42.—Graph to show how the frequency of a character varies with the frequency of the genes (see text).

panmixia. If panmixia occurs, it is easy to carry out calculations about the distribution of simple recessive and dominant characters (Fig. 42).

For instance, let us deal with a single recessive gene

Race, Reason and Rubbish

R, calling the corresponding dominant D instead of using the previous convention of representing a recessive gene by a small letter. The combinations which can then turn up are: (*a*) RR, i.e. persons with the recessive gene in duplicate and therefore with the recessive character; (*b*) DD, i.e. persons with the dominant gene in duplicate and therefore with the dominant character; (*c*) DR, i.e. persons who have only one dominant gene and carry the recessive one, though they show the dominant character. If among all the pairs of such genes present in the population one-tenth (0·1) are R and nine-tenths (0·9) D, the following combinations will exist.

$$
\begin{aligned}
RR &= 0\cdot1 \times 0\cdot1 &&= 1\% \\
RD &= 2 \times 0\cdot1 \times 0\cdot9 &&= 18\% \\
DD &= 0\cdot9 \times 0\cdot1 &&= 81\%
\end{aligned}
$$

It is easy to demonstrate that these figures are correct. We will assume that we are dealing with a game of chance, and that we get the same sort of results as we should get if we were playing some sort of lottery with a one in ten chance of winning (i.e. getting an R), and a nine in ten chance of losing (getting a D). If we want to get two R's in succession we only take a second draw after getting an R at the first. If we drew one hundred times we should get ten R's on the average in the first draw. The question of going on to a second draw therefore arises only on these ten occasions, and only in one out of ten of these shall we get a second R.

Race, Reason and Rubbish

Thus the combination RR turns up in one out of one hundred, i.e. in 1% cases of a double draw. In the same way we can reason about DD. If we draw one hundred times we get D on ninety occasions. Only on these ninety can we go on to get a double, and in nine-tenths of these ninety we get a second D, i.e. in 81% of all double draws on the whole.

It is a little more complicated to explain the probability of getting RD. We must here remember that we have now two possibilities, i.e. to get R to begin with and D afterwards, or get D at the first draw and R at the second. Thus if we ask what is the probability of first getting an R, the answer is ten cases in one hundred. We can draw again after getting these ten, and in nine out of ten of them, i.e. in nine out of the original hundred, we then get the combination RD. At a first draw, however, we got ninety D's, and when we draw again we could get the combination RD in one-tenth of them, i.e. nine out of the original hundred. Altogether we thus get RD or DR in 9 + 9 cases out of a hundred, i.e. 18%, and if we add all our percentages together the sum is 1 + 81 + 18 = 100. This is a check on the accuracy of our calculations, and a check that we have taken into account all the possibilities imaginable.

Calculations of this sort can be carried out with any proportions whatever. So if we know the number of individuals who have a recessive character, we can estimate in how many of the population the recessive

Race, Reason and Rubbish

gene is latent. Fig. 42 shows a diagram of the different possibilities. To get the percentages, we read off the vertical distances between the graphs. We see from this diagram that the maximum possible number of latent carriers of the gene is 50 %. Individuals who exhibit the recessive characters then make up 25 % of the population, as also do homozygous dominants. We then have just the same proportions which we should get according to Mendel when we cross a pair of heterozygotes, i.e. one-fourth recessives and three-fourths individuals with the dominant character. Two-thirds of the latter have a latent gene.

It is interesting to make a similar calculation for a much more rare gene. Let us assume that only 1 % of all the genes of a particular pair in the population as a whole are recessive. By the method used previously we find that when this is so the recessive character will occur in $\frac{1}{10} \times \frac{1}{10} = 0.10$ % of the population. The gene is latent in $2 \times \frac{1}{10} \times \frac{99}{100} = 1.98$ % of the population. This calculation brings out a feature of fundamental interest. We had already found that if the recessive character occurs in 1 % the gene is latent in eighteen times as many individuals, i.e. 18 % of the population. We now find that if the character is present in 0.01 %, a single dose of the gene is present in one hundred and ninety-eight as many individuals, being latent in 1.98 %. It is possible to show that these figures are representative of a general rule. As a gene

Race, Reason and Rubbish

or a character becomes more rare, the proportion of individuals in which the gene is latent becomes greater in comparison with those in which the character is manifest. This is really quite easy to grasp. As a rule a very rare gene should be present in single dosage, and only very exceptionally in duplicate. If, on the other hand, it is common, and especially if it is very common, it should often turn up in double dosage, and cannot therefore occur among so many individuals in a single dose. Such calculations, and the rule based on them, are not merely of interest as curiosities. They can lead us to conclusions of practical importance.

We shall come back to this question. Meantime, perhaps some of us will wonder whether there is any reason for assuming a random process of this description. The fact is that everybody who is married likes to feel that his or her own marriage was not a matter of pure chance. All the same, it is quite clear that only chance decides whether the contracting parties have one or another *latent* gene; and we must also admit that choice occurs at random with respect to many actual characteristics, such as blood-groups or colour-blindness, as for any latent morbid gene. So far as many characters are concerned, we are therefore quite right in assuming that *panmixia* takes place. We can show this by an actual example. Women (see p. 109) become colour-blind only when they get the gene in double dose, but men exhibit the character when they have a single dose

Race, Reason and Rubbish

of the gene. They cannot have it in duplicate because it is sex-linked. Colour-blindness occurs among roughly 8% of men, and the figures do not differ much in different populations. Starting from this figure, which may be taken as approximately correct, we should expect to find colour-blindness in $\frac{8}{100} \times \frac{8}{100} = 0.64\%$ of women. Statistics about women have been collected and agree with this estimate.

Even if we are entitled to assume that *panmixia* occurs comparatively often, it is also certain that discrepancies exist. There are five possible ways in which discrepancies can arise.

1. A man or woman can avoid marrying or have less than the average number of children. If this happens on a large scale we call it *selection*. In this way certain genes have less chance of passing over to the next generation.

2. The individuals concerned can marry relatives, and if this occurs more often than by mere chance we call it *inbreeding*.

3. We can choose our partners on account of particular characteristics. Probably, for instance, musical individuals marry one another more often than they marry persons with no such gifts. This is called *assortative mating*.

4. When people choose partners they do not have access to the whole population. We marry within the confines of a circle, which is limited by geographical

Race, Reason and Rubbish

boundaries. Thus a person in Reno has not so much likelihood of marrying a person in Boston as of marrying another person in the same resort. Besides this, there are several reasons why people marry within a class or "racial" group. These limitations are collectively called the *isolate effect*.

5. The gene equipment of an individual can undergo change, because new genes turn up. This is called *mutation*.

So far as the author can see, there can scarcely be any other exceptions to the rule that mating really does occur at random. In so far as this happens, it has now been shown that the hereditary equipment of an individual remains constant from generation to generation. Our next job is to try to disentangle changes which may occur on account of the five processes mentioned above. We shall now discuss each of these problems in turn.

CHAPTER VII

WHEN we have talked about a population with a stable gene equipment maintained without change in successive generations by fixed fertility and mating at random, nothing said so far need imply denial of the possibility of fortuitous changes. The frequency of a gene can increase to some extent in one generation and diminish in another; but changes of this sort must cancel out in large populations. It is only in small communities that we have to reckon with appreciable deviations in successive generations. Similarly, when we speak of people marrying at random and producing an average number of children, we do not exclude the possibility that there are many children in one family and few or none in another; but if the average rate of reproduction in a particular group of individuals with a particular equipment of genes is higher or lower than it is in the community as a whole, successive generations will be different. What then concerns us is to get a more precise measure of how quickly such changes take place, i.e. to try to estimate how effectively interference with reproduction produces results. If processes of this kind go on in a population, we speak of *selection*.

When people who have a *dominant* character are completely prevented from reproducing their kind, the gene, and therewith the character, is instantaneously

Race, Reason and Rubbish

eliminated. This is the explanation of why no severe hereditary diseases, contracted early in life, are inherited as dominants. Persons who suffer from such a disease, for instance an extreme type of idiocy, rarely get children, if at all. In a way this means that the disease destroys itself. Even if occasional reproduction does occur, the character will soon disappear. Suppose that it corresponds to one-tenth of the average fertility in the population. Those who have the character in one generation will be one-tenth as numerous as in the previous one. In the next generation there will be one-hundreth, in the next but one one-thousandth, and so on. Even a moderate difference of fertility will quickly produce a depressive effect on the frequency of a dominant character. If, on the other hand, those who have a dominant characteristic have somewhat more than the average number of children, and those with the corresponding recessive one have less or none, the mechanism of change must be different, because the recessive gene will be latent among a section of individuals with the dominant character; and such individuals do not encounter any impediments to reproduction.

We now know (p. 144) that, if a recessive character is common, persons in which the gene is latent are relatively less common than if it is rare. For this reason it is easy to see that selection acts in a comparatively effective way on a recessive character which is common, and that it is less effective when it is rare. If we make precise calculations on the assumption that all who have the character are prevented from propagation, we can make a graph of the changes. The graph falls steeply

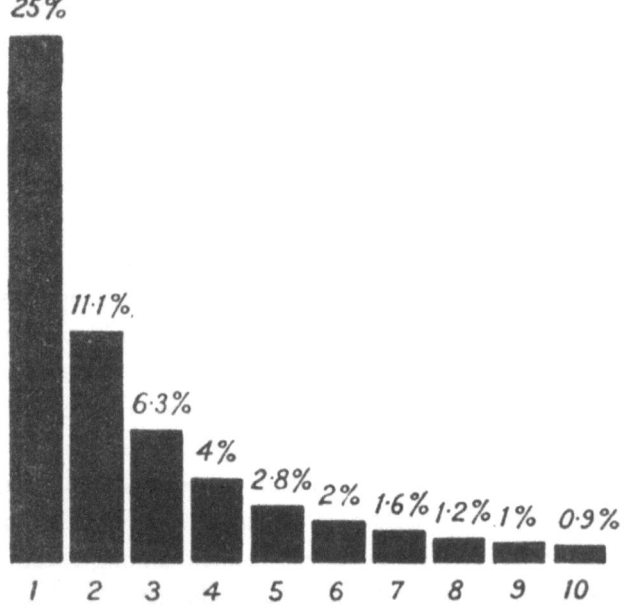

FIG. 43.—The Effect of Selection. The heights correspond to frequency of recessives in successive generations, when all recessive individuals in each generation are prevented from having any offspring

at first, and thereafter approximates more and more slowly to the horizontal (Fig. 43). For instance, we find that if a recessive character is present in 25 % of the population, completely effective sterilisation of all who have it diminishes its occurrence to 11·11 % in

Race, Reason and Rubbish

one generation. Gradually in this way it reaches a comparatively low percentage (Fig. 43), while the effect of sterilisation becomes weaker all the time. If a recessive character occurs among 0·1 % or ten in ten thousand, we should need to continue with complete sterilisation for ten generations, i.e. about three hundred years, in order to diminish the frequency to 0·06 % or six in ten thousand. The situation is therefore as follows. Adverse selection of dominant characters is always effective, and we can always rely on getting a quick effect when selection is directed against *common* recessive characters. The same is true, but to a less extent, of characters transmitted in a more complicated way, i.e. those which result when several genes of dissimilar pairs get together.

This aspect of the matter is interesting because the differences of fertility which distinguish contemporary social classes and occupational groups would have an effect by no means negligible, if such groups had very different assemblages of genes. One thing much discussed in this connexion is the low fertility of the prosperous classes. Some say that this low birth rate constitutes a danger to society. They assume that the upper classes are more especially the carriers of genes which promote social effectiveness, intelligence, industry, or high character. The same people often complain that the low fertility of well-connected people points to an egotistic or socially indifferent attitude, and that it is therefore

Race, Reason and Rubbish

essential to encourage a more idealistic outlook in connexion with production of children. It seems rather odd to make out that the upper classes are too selfish, while also urging that such selfish people should be specially encouraged to reproduce.

On the other hand, it may also be said that having few children is not necessarily due to pronounced egotism but to a stronger sense of responsibility for offspring; and that people may not wish to have large families because of the uncertainty of providing for many children so well as they could provide for a few. Even if they are able to help a large family better than other folks, prudent people may feel convinced that they can look after one or two children better than six or seven. On this statement of the case we may perhaps exonerate the prosperous classes from the charge of exaggerated egotism, and what we have then to decide is whether they really constitute a selection of specially intelligent people. Other things being equal, it is quite clear that intelligent people have greater possibilities of getting on in the world than stupid ones have; but many people who belong to the upper classes have not worked their way up to the positions they hold. They inherit them. Although their ancestors may have done the necessary work for rising in the social scale, it is not at all certain that the offspring of such individuals constitute as good a selection as their forefathers.

We may also ask if the process of social selection is

Race, Reason and Rubbish

itself a good one, i.e. whether people who have worked their way up in the world are really over-blessed with desirable characteristics. As emphasised already, intelligence must be some advantage; but many other traits also play a part in the individual struggle for social position. A certain lack of scrupulousness may help. The very lack of intelligence can lead to equanimity in critical situations and may therefore propel people slowly upwards in some circumstances. For instance, those who lead the crowd and carry people along with them by force of conviction are seldom intelligent. To make an impression on the masses calls for the assurance that personal opinions are correct. Only those who have this assurance can talk with enough conviction to get results. Only the stupid can arrive at complete certainty. Only those who are not much above the average can express themselves in a way which the masses understand.

Needless to say, the nature of social processes which determine class mobility are too little understood to entitle us to give a decisive verdict. A number of enquiries have been carried out by means of so-called intelligence tests upon children of different social classes. The differences discovered are comparatively insignificant, and only lead to the conclusion that the average divergence between different social classes is not of very decisive dimensions. With regard to inheritance in general it is important to recognise that traits which play a part in social efficiency are very largely influenced by the social milieu itself. They depend partly on early training, school education, and so forth, and we may be sure that the hereditary mechanisms also involved are of a very complicated type. Whatever we may mean by the word intelligence, it is certainly not inherited as a simple Mendelian character, but depends on a multiplicity of genes. Both these considerations diminish the likelihood that big differences of gene equipment distinguish different social classes.

Let us also remember that possibilities of getting on socially—especially in earlier times—have been decidedly limited. We can get a picture of the real state of affairs from Fig. 44, which shows percentages of successive age groups in the Swedish school-leaving examination. Though there are some opportunities for a career without the certificate, they are very limited. So it is the ticket of admittance to the more socially esteemed occupations. It can be seen from the picture that a comparatively small part of any year-group take the examination. Previously it was mainly children of the well-to-do who had the opportunity for doing so. Of late the number of students has increased. A process of democratisation has gone on. Far more than previously, the school-leaving examination is now taken by people whose parents belong to the lower middle and to the labouring classes. This is mainly so in town populations. In the country it is still chiefly the children of the well-

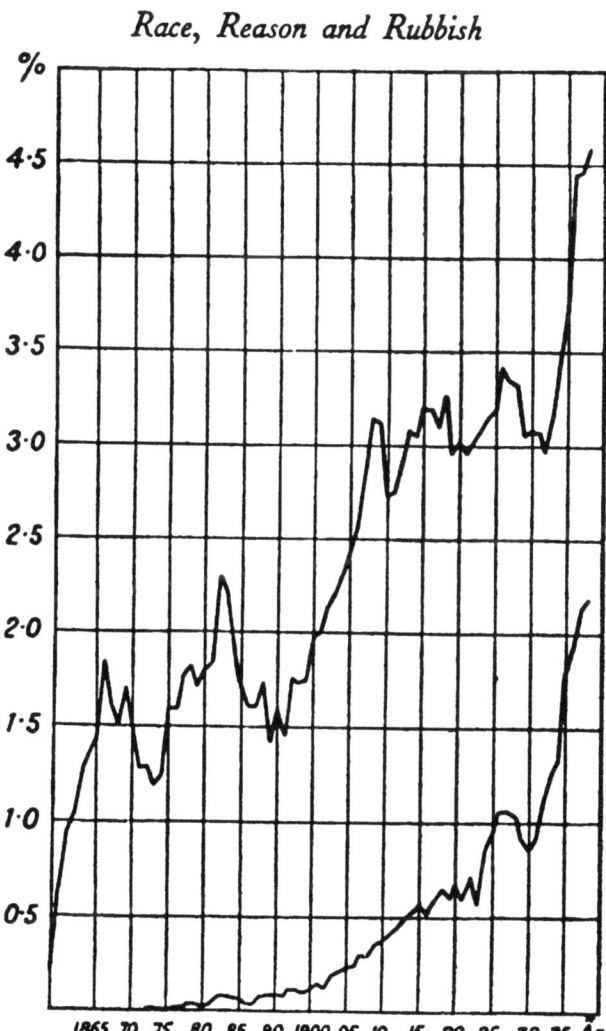

FIG. 44.—Graph showing percentages of each year-group for the school leaving examination (p. 154). Males (above) and females (below) are plotted separately

Race, Reason and Rubbish

to-do who have opportunities. On the basis of the figures set forth in the diagram, we can conclude that the majority of annual recruits to the more prosperous classes come from the same social stratum as their parents, and that only a small number of gifted individuals of lower social position have the chance of rising.

Beyond this we can scarcely go at present. We must carry out comprehensive researches before we can venture to give a more definite verdict about this important question. Meanwhile we should remember that differences of fertility such as those we have discussed are of comparatively recent date. In France they have existed for a few generations; but in most other countries they are scarcely more than a generation old. Besides, we have reason to believe that family limitation began among the well-to-do and spread to the poorer classes. In Stockholm, and in some other parts of the world, people with low incomes have begun to have less children than those with larger ones. So it is likely that such differences between social classes are ephemeral. As the problem of a declining birth rate grows and assumes importance, it is probable that differences of fertility between people of different income levels, such as those we have discussed, will disappear.

In so far as it affects smaller groups, the problem of selection is comparatively simple. If a rare character associated with low fertility, or one which is inherently associated with obstacles to reproduction, is inherited as a dominant, it will soon be eliminated. Thus the effect we get from sterilisation is proportionate to the number sterilised. For example, if we sterilise half of all people who have a particular dominant character, the occurrence of the character is diminished by one half in the next generation. On the other hand, where we have to deal with recessive characters or traits transmitted in a more complicated way, the results of interfering with reproduction are without practical significance when the character itself is a rare one. This assertion contradicts widely accepted opinion. People have worked for sterilisation measures with the more or less explicit aim of diminishing the number of defectives in a population.

Perhaps it is not convincing enough to carry out merely theoretical calculations about the effectiveness of selection, as in what has gone before. We do not feel convinced by arithmetic if we cannot control it in practice. So it is reasonable that we should want more tangible evidence. In this connexion the author has emphasised that we already have such evidence in the occurrence of juvenile amaurotic idiocy. This is a form of idiocy combined with blindness. The unfortunates who inherit it die about the age of puberty and never have children of their own. About four or five such individuals are born annually in Sweden, and there are between seventy and a hundred of them alive at the

Race, Reason and Rubbish

moment. Though this character is subject to so-called natural selection more exacting and stern than could be brought about by any law of sterilisation, and though selection operates wherever the character exists, the disease has not been eliminated. This may seem a little odd. We should expect that prevention of reproduction among such individuals would have some influence. We should certainly anticipate that it would lead to the total disappearance of the character in the course of millennia. In reality this is not so. The effect is so insignificant that it would not be noticeable in the comparatively short interval of time which corresponds to the duration of human life on earth.

Let us look at the matter in another way. The gene for a recessive defect must have turned up at least once through a mutation. Perhaps the first person who has the gene does not have children. So the gene disappears. It may also happen that he has several children of whom some get a single dose of it. This allows the gene to be transmitted to their offspring, and it may gradually spread throughout the population. When the gene has spread to some extent in this way, two persons in whom it is latent may happen to marry one another. Among their children we should expect that one of every four will actually exhibit the recessive character, two others will carry the gene, and a fourth will lack it altogether. Now if we prevent those who have the character from reproducing their kind, we cannot reasonably hinder

Race, Reason and Rubbish

both of those who carry it, and if all families consisted of four children and two parents, there would be two carriers among the offspring for every two parents also carriers. So by getting rid of those who have the manifest character, we do not necessarily affect the number of those in which the gene is latent. A recessive gene must have a certain spread before characters begin to crop up at isolated points in the population. Even if we capture these points we can never suppress the gene below the level of frequency which corresponds to that at which this isolated outcropping of the character occurs. This lower limit is what the author of this book has called the *lowest heterozygote limit*.

In this connexion it is worth rubbing in the fact that corresponding reasoning applies even to defects with a more complicated method of transmission, for instance characters which depend on two dominant genes which must come together if it is to show up. Even if we adopt measures for the sterilisation of parents and brothers or sisters of every defective at its birth, we could not change the situation. Rare genes could not be completely stamped out. They can only be pressed down to a certain limit of rarity.

Hereditary defects and inherited diseases are comparatively rare phenomena. Even if we set the limit at its highest, we need scarcely reckon with more than a few per cent of really defective people in any population. An estimate by the author for the Swedish population

Race, Reason and Rubbish

gives 4%, even when the limit is stretched to the uttermost. If we lump together all defective individuals, we are perhaps tempted to believe that we could produce a decided diminution because the figure, though low, is not so low as assumed in the reasoning set forth above. However, we have to remember that the group is not homogeneous. It is made up of numerous dissimilar diseases such as hereditary blindness, deafness, congenital disabilities, epilepsy, idiocy, mental dullness, etc. With each of these classes in its turn we may have to reckon with several different types of transmission. Among the blind, for example, we can distinguish at least ten types, and there are many quite different sorts of idiocy or imbecility. For every such sub-group the effect of sterilisation is zero, and the sum of a set of zeros is also zero. In other words, all these defects lie near to the lowest limit of heterozygote frequency.

If we doubt whether this conclusion is a reasonable one, we ought to take into account the fact that such defective individuals have had little opportunity of reproducing during the last few thousand years. They have lived on the verge of hunger and death, dependent on charity which scarcely existed. They could not procure their own means of subsistence; and we may judge the prospect of life for them from the fact that the custom of exposing the newly born with any obvious defect persisted till a comparatively high level of culture. Though people could not bring themselves to kill

Race, Reason and Rubbish

helpless children, they let them die by themselves of cold and hunger, and surely did so with no lightness of heart. More human sentiments are of comparatively recent date; and still later came kindly treatment of the children of social misfits. In such circumstances the reproduction of defective individuals must have been practically non-existent in the past; and it is certainly very small at the present day. Despite stringent selection, defectives have not been stamped out. So far as we can form a rough opinion, for which there are no exact figures, they exist also among peoples whose social attitude towards the care of the helpless is not highly developed. Even in primitive communities blind, deaf, and crippled offspring are born.

Though we cannot expect to produce any socially significant reduction of the frequency of defects by measures for sterilisation, there is no reason why we should not take *individual* precautions when well-founded conclusions lead us to believe that we can prevent the birth of an individual whose life will be a burden because of some blemish or the like. In individual cases we can prevent a misfortune, and since defective individuals may be a quite appreciable economic burden to a small community, there is no reason why we should not do what we can do for the individual, even if it is not very significant from the standpoint of the population as a whole.

The conclusion to which we have come has also a

brighter side. Since we cannot stamp out rare defects, we cannot stamp out advantageous characters which are rare, especially when inherited in a complicated way. For instance, we cannot get rid of intelligence. People have actually tried it on. Celibacy in the Roman Catholic Church is a trick of this kind. At one time, more gifted individuals chose to take vows, and for many gifted individuals of the poorer classes such a decision was the only opportunity of drinking from the springs of culture. Though celibates were prevented from having children, Roman Catholic countries have not any outstanding lack of gifted individuals. Even if some countries adopt measures specially directed against persons who have sufficient intelligence to form independent opinions in opposition to prevailing superstition, we need not expect that they will be able to prevent the birth of single individuals with more than normal intelligence. Stupid people cannot stamp out genius.

CHAPTER VIII

PAIRING between brothers and sisters or between parent and child occurs among animals, though some stockbreeders allege that there are difficulties in getting a bull to pair with its sisters. With regard to human beings we may take it that connexions of this sort are forbidden in practically all communities. The reason for this has given rise to question, and people have chosen to see an intelligible biological explanation for the disinclination of our own species for close inbreeding. Some say that it brings about degeneration, others that people have thus acquired an instinctive repugnance to it. It may also be relevant that near relationship brings about an intimate mutual understanding of shortcomings and defects, and that such knowledge is inimical to the production of an illusion which is often a prerequisite for marital union. As a matter of fact, intensive inbreeding does not necessarily bring about harmful results among human beings. Cleopatra was the product of several generations of brother-sister mating.

Research on inheritance has now provided us with solid ground for analysis of the problem of inbreeding. Inbreeding brings about a reduction of the proportion of heterozygotes and an increase of homozygotes in a population. By brother-sister mating for successive generations, like genes are increasingly brought together in duplicate, while individuals with a latent gene become more and more rare. If we want to get pure lines and individuals which have reliable offspring to-day, we therefore use inbreeding extensively. It is a fact that all populations harbour a large group of genes for defective conditions transmitted as recessive characteristics. The really bad defects could not well be dominant (see pp. 148–149). So if we practise inbreeding we get an increased number of defective individuals. This is the germ of truth behind the belief that inbreeding produces degeneration; but it is also true that we gradually reach an equilibrium if we continue inbreeding. A fixed proportion of types is then maintained without further degeneration. Thus the proportion of defective individuals no longer gets bigger.

On the basis of experience with animals and plants it used to be emphasised that inbreeding can also act advantageously or disadvantageously so far as we ourselves are concerned. A cousin marriage in a family with good characteristics is advantageous; but if there are bad traits in the pedigree, as for instance cases of diseases which are inherited as recessives, a cousin marriage will be ill-advised. The problem which interests us here is not the isolated case but how consanguineous wedlock, i.e. marriage between relatives, affects the gene-equipment of the population as a whole. For a population in its entirety inbreeding should act in accordance with the principles which apply to

Race, Reason and Rubbish

animals and plants. Homozygous individuals must increase, and the occurrence of individuals in whom the recessive gene is latent, or the dominant one in single dosage, must diminish. But the question is not disposed of by this bare statement or principle. We must try to get an idea of how powerful this effect may be.

To all intents and purposes crosses between parent and child or brother and sister do not happen. Marriages between uncle and niece or aunt and nephew are very rare, and in many countries they are forbidden. So the type of consanguineous union which is likely to be at all significant is first-cousin marriage. To get an idea of how significant it may be, we can calculate the result of 100 % cousin marriage. This means that we can assume that all marriages in a population are marriages between cousins, and then compare the result we arrive at with what happens when mating is haphazard. If a recessive character is present in 25 % of the individuals mating at random, 100 % cousin marriage would raise the proportion of recessive homozygotes to 26·6 % in one generation. The difference of 1·6 % is by no means a big one. If the character has a frequency of 9 % it could be raised to 10·3 %—an increase of 1·3 %. If we consider a character which is so rare that it occurs in only 1 % of the population, the result we get is 1·6 % and thus an increase of 0·6 %.

So by comparison with the powerful effects which

Race, Reason and Rubbish

inbreeding can produce on animals, any *practicable* result which could be got with human beings is insignificant. In reality, it is indefinitely small, because cousin marriages are comparatively rare phenomena and make up nothing like 100 % of all marriages. The extent of cousin marriage has varied at different times in different countries, but such unions have seldom exceeded 1 %. In civilised countries of Western Europe nowadays the numbers lie round about 0·5 %. On the assumption that no appreciable amount of inbreeding results from pure chance, this means that the contribution of cousin marriages to human populations corresponds to about one two hundredth of the contribution of mating at random.

We must remember that the assumption of marriage at random implies that unions of cousins may sometimes occur. For practical purposes we may regard the chance occurrence of such marriages as zero when a population is very large. In reality populations are not generally so large. As intimated already, any population can be split up into groups within which we can speak of random mating with some justice. People who live in the country marry within small districts, and people who live in large towns marry within wider social circles. The frequency of cousin marriage is actually higher in the country than it is in the towns. So presumably cousin marriage occurs in roughly the proportion of cases prescribed by pure chance.

Race, Reason and Rubbish

Let us now take a step further, and assume that cousin marriage is completely prohibited. Thus we reduce the frequency of cousin unions from what we should expect if mating occurs at random to zero. Even this has no significant effect from the standpoint of the population at large. Whether this assertion is justified is a matter for discussion when we are concerned with rare characters. Everything depends on how big a change we care to call an effect. The real state of affairs is that, though really insignificant from the standpoint of the population, the effect of cousin marriages need not be entirely negligible, if we look at the matter from the standpoint of the individuals who exhibit a recessive trait. If a recessive trait is common, a greater or smaller amount of inbreeding makes little difference to its occurrence. If it is less common, the effect is somewhat greater though still insignificant from the standpoint of the population itself. If it is extremely rare, inbreeding can greatly increase the occurrence of the character, but because the latter is extremely rare a *relative* increase has no appreciable mass effect.

To get the situation clearer we must go further into the problem with special reference to very rare characters. Let us suppose that a gene arises in some population through mutation. If recessive, it cannot make itself felt before it has spread out in successive generations. Eventually two persons who carry the gene may marry. Two such individuals are descended from the one who had

Race, Reason and Rubbish

the gene in the first place. So the character can come out only through consanguineous union, and at earliest by cousin marriage. This means that if a gene turns up in one person, it must hide itself for at least three generations, i.e. roughly a century. It means also that cousin marriages have a comparatively great significance for extremely rare characters.

We now need to get some measure of how great this can be; and we can put the issue in a practical way by asking how much we can diminish the occurrence of a rare character, such as a congenital defect, by forbidding cousin marriages. If we could put a stop to a large number of cousin marriages, we could get a big result; but we can hardly imagine any real population in which there are many cousin marriages to prevent and at the same time any rare characteristics. In real life we may have to deal with large population groups, such as a large city, where really rare genes occur, but if so there are relatively few cousin marriages to stop. If we are concerned with a small one, many cousin marriages occur, but in a small community no gene can remain really rare. This way of looking at the problem may seem difficult at first sight, but not so difficult if we bear in mind the fact that when we speak of a *very rare* character, we usually mean a trait such as *albinism*. According to a British estimate, this has a frequency of about one in twenty thousand. For a recessive gene to show up, the population must contain at least two

Race, Reason and Rubbish

people in which it is latent; and only when two such people marry and have children can we get individuals who show the recessive character because they have the recessive gene in duplicate. In a small community of fifty to one hundred individuals the existence of only two carriers does not constitute a very rare phenomenon in the sense defined.

Calculations which are relevant to this question are somewhat complicated, but the upshot is that the degree of rarity of the gene on the one hand, and of cousin marriages on the other, balance one another. So at best we can only reduce the incidence of a simple recessive character by about 15 % if we put a stop to all unions between cousins. Since all such theoretical estimates call for a check, it is proper to ask whether this conclusion can be verified. It transpires that about 15 % of parents of persons who suffer from extremely rare diseases are actually first cousins. We have not much material at our disposal and figures vary at random a little. For instance, the figure for a nervous disease called *Friedreich's ataxia* is 9 %, for an eye disease called *retinitis pigmentosa* it is 17 %, and for *juvenile amaurotic idiocy* 15 %.

To some extent it is a matter of taste whether we consider an effect of this order of magnitude significant or trivial. The reduction merely applies to very rare characters, and individually such characters have no significance for the population at large, but, as emphasised in an earlier chapter, there is a fairly large collection of recessive genes for defects in any community. So perhaps a figure as large as 15% should not be treated as entirely negligible when we look at the sum of all the defects which depend on such genes. At the same time we have to remember this. Though prohibition of cousin marriages leads to such a result in the next generation, it does not produce a further diminution in succeeding ones. The result is instantaneous; but there is no continuous gain from one generation to another. Where we have to deal with characters which are not so rare, and are not therefore near the *lowest limit of heterozygote frequency* (p. 159), the effect obtained is less considerable. The effect is also less considerable when a character is transmitted in a more complicated way.

So far, we have only discussed single recessive genes and traits inherited in a complicated way. When dominant characters are concerned the picture is reversed. Traits determined by recessive genes increase a little because of inbreeding. Traits which depend on dominant ones must diminish. By stopping consanguineous marriages, we spread the dominant character, because we prevent two dominant genes getting together in the same individual. None the less, this result is quite insignificant. Indeed it is not a thing to be reckoned with in any circumstances. If the gene is very rare it is generally found in single dosage. If all individuals who have a single dose of it also exhibit the character, prevention of cousin marriage, and hence the appearance of a few more homozygotes, signifies nothing from the standpoint of the population or even from that of individuals who possess the trait.

With reference to inbreeding in human populations the outcome of our enquiry is therefore this. Neither the grade of consanguineous marriages nor the percentage of those which actually occur can be said to have any noteworthy significance. Only when very rare recessive characters are concerned, or when the method of inheritance is complicated, can the exclusion of cousin marriages have an appreciable effect. Perhaps the reader will ask at this point if it is so great that we really ought to take action by preventing cousin marriage. Before trying to answer this we should remember that we cannot rest content with forbidding some cousins to marry. We have to stop all of them if we hope to get measurable consequences, and we do not in fact know whether a gene is latent in any particular pair of cousins. It is therefore a matter of settling whether healthy people should be compelled to make sacrifices in order to get a diminished frequency of defectives in a later generation. Obviously it is not the province of science to decide how many of us should immolate ourselves on the altar of the coming generation.

How much the misfortune which prohibition of cousin marriage would entail for particular individuals is compensated by the advantage which a diminished frequency of defectives signifies cannot be assayed. To discuss the matter in detail we need more reliable figures about the frequency of defects than those at our disposal. So the issue at present is not a real one. Meanwhile, the author of this book regards the possibility with misgiving. For the time being, in many quarters, inclination to apply the results of race biology in practical policy is too strong rather than too weak. People do not demand the certainty which they ought to seek before they undertake thorough-going reforms.

CHAPTER IX

RANDOM mating means that people who have particular characters marry one another as chance dictates. If such marriages take place more often than chance dictates we speak of *assortative mating*, i.e. choice of *like for like* or the converse. Choice of like for like must work in much the same way as inbreeding. That is to say, it leads to a diminution of the number of heterozygotes and an increase of the number of homozygous individuals in a population. If everyone with a particular trait married another person with the same one, the effect would be very considerable. It would lead to the disappearance of heterozygotes comparatively soon. For instance, when it is a question of a relatively rare character, choice of partners first leads to doubling its frequency in each generation. Little by little the increase becomes smaller. Eventually we get to a state of balance. To all intents and purposes no carriers of the latent gene then remain in the population.

We may well ask ourselves whether it is reasonable to suppose that anything of this kind happens in human populations. Consistent choice of like for like can scarcely be characteristic of any marriage, but to some extent there must certainly be selection of this sort. For instance, we know that tall people marry one another more often than mere chance accounts for. Clearly, the

Race, Reason and Rubbish

fact that a tall and a short person may produce a slightly comic effect when in company together is not an entirely trivial consideration where marriage is concerned. We may also presume that musical people quite often marry one another.

The most extreme case of assortative mating is surely among deaf people. Deaf-mutes have special connexions which promote meeting together. For a deaf-mute the difficulty of making the sort of contact which leads to marriage with anyone who has not the same disability is surely great enough to make wedlock between themselves and normal people an unusual occurrence. On the other hand, the mischance is itself a bond of union between those who share it, and with the help of special organisations which exist for their benefit can be assumed to bring about relatively strong assortative union. At present no quantitative researches have been undertaken in this connexion.

How far a corresponding process occurs among persons with outstanding intelligence has not been cleared up. It goes without saying that an intelligent person does not generally like to marry a stupid one. On the other hand, good looks, social standing, and accidental circumstances of various kinds, have so much to say that we could not make any confident statement without special enquiry into the facts of the case. Meanwhile there is insufficient reason to believe that good looks have any special connexion with brains, though

Race, Reason and Rubbish

one investigation into the connexion between school performance and personal appearance has been made in America. Girls in a high school were classified according to looks as: handsome, sweet, quite passable, and homely. A detectable but not pronounced parallelism between these grades of personal appearance and quality of school work was found to exist. On the whole, those who had the advantage so far as looks are concerned were not so good at their lessons as those who were less easy on the eyes. The difference may conceivably be explained by unequal attention to studies owing to male appreciation of more superficial merits.

While it is clear that choice of like for like may in some circumstances exceed anticipations based on the assumption that marriage is a lottery, it is not so clear that the converse is true. There is a popular belief that partners to a marriage may be drawn together by dissimilarity of temperament. Though there are no available data to prove that the attraction of opposites is a common occurrence, we cannot exclude the logical possibility that it affects the distinction of some characters, especially with respect to emotional make-up. The effect of *negative* assortative mating, to use its technical name, must be the reverse of what occurs when there is choice of like for like. It would diminish the association of homozygous types and increase the proportion of heterozygotes. When recessive characters are involved, the effect of negative assortative mating is therefore like

Race, Reason and Rubbish

that of adverse selection. If a recessive character is common its frequency may be greatly diminished, but there will be no appreciable result if it is rare. From a theoretical point of view this possibility is specially interesting because it would still check the tendency of a community to split up into biological castes, even if there were no economic or social circumstances operating in the same way.

As a matter of fact, there are scarcely any recorded figures about the way in which selection of partners occurs. So we cannot reach definite conclusions at present. One special drawback is that the mathematical treatment of the problem has not been fully explored. We can calculate the effect of 100 % assortative mating, but we do not at present know what effect different intensities may also have. One reason for this is the difficulty of making plausible assumptions in this field. A fictitious example will show up some of the difficulties. Let us assume that musical persons, while not always marrying one another, do so twice as often as chance prescribes, and let us make the further assumption that musical ability is transmitted as a recessive character. The proportion of musical individuals in the population is thereby increased. We gradually reach a higher percentage, and we can ask, what must now be the frequency of marriage between musical people, if they continue to have the same disposition to enter into marriage with one another—and to the same extent? Obviously,

Race, Reason and Rubbish

musical people cannot always marry one another twice as often as mere chance dictates. If the percentage in the population increased above fifty, such doubling would be an arithmetical impossibility.

For the present there is little more to be said about the problem. There is only one detail which should perhaps be emphasised. This is the fact that the attraction of like for like can scarcely take place when the characteristics involved are of the rarer kinds. Persons who have such special characteristics are too infrequent to be able to meet one another to a greater extent than chance dictates. Each moves in his or her own circle in his or her own isolate. In exceptional circumstances two such may marry by chance; but this cannot amount to much, unless special conditions prevail, as in the case of deaf-mutes. If positive assortative mating does prevail, and has done so for a very long time, a state of balance would have now been reached. Meanwhile displacements of social arrangements, which depend on the evolution of transport, immigration to the towns, and so forth, entail an increased possibility of marriage between individuals with similar characteristics. So we have some reason to believe that the effect of assortative mating had reached a state of stable balance in recent times, but choice of like for like has lately increased, and now plays a more important role. We are not justified in making a definite statement about how big the effect of this is.

Race, Reason and Rubbish

Although the genetic equipment of the population and the proportions of different types remain constant when mating occurs at random, and no differential fertility exists, changes do come about, because new genes turn up. Experience we gain from the world of plants and animals shows that *mutations* (p. 87) are not very common. We have good reason to believe that inherited diseases have turned up in this way, and without doubt genes which determine such conditions are still doing so, but if we confine ourselves to discussing the evolution of mankind during the next few generations, i.e. one or two centuries ahead, we have little reason for believing that mutations will have any noteworthy influence on the course of history. The outcrop of new characters by mutation is significant when we are concerned with long periods of time, and when we are looking for an explanation of the appearance of new species and evolution towards higher, i.e. more complex, organic types. Even here we have not yet got a completely clear view of the problem.

It is scarcely reasonable to embark on a serious discussion of the future of the human family several generations ahead. Its prospect is largely dependent on scientific discoveries and technical possibilities which are moulded by them. The difference between social arrangements ten generations ago, that is to say, three hundred years before our time, is so great that people would obviously have come to a completely erroneous

Race, Reason and Rubbish

picture of the present, if they had then tried to make prophecies about future social conditions. Inevitably we try to peep into the future, and, inescapably, we try to estimate the ultimate effect of contemporary occurrences and processes. But we cannot go beyond making a more or less likely forecast of the immediate future. When longer periods of time are involved, the only certainty is that all our forecasts will be faulty. On the other hand, it is quite clear that as we get deeper knowledge, the forecasts we can make will be more reliable. Needless to say, any statements about the hereditary make-up of populations made in this and other chapters therefore apply only to social conditions which now prevail. If social organisation changes fundamentally, the situation may be changed in many ways, even with respect to processes which involve inheritance, as here discussed. Assertions made are naturally valid in so far, and only in so far, as the conditions assumed hold good.

CHAPTER X

In what has gone before it has been emphasised repeatedly that a population consists of groups within which marriage is more or less a matter of chance. Such groups, called *isolates*, are certainly of very different dimensions. An isolate may be a small out-of-the-way village which has bad communications with the outside world. At the opposite extreme is the large city with amenities of public entertainment. There, marriage may take place haphazard within the confines of a much larger group. Boundaries between isolates are partly geographical. Forests and seas, or rivers and mountains, cut off districts from one another, and evidently more so in former times than to-day. There are also frontiers of a different kind. A person marries more often within his or her own class than with others at a different social level. In this sense we can speak of *social* isolates. One type of such isolates are certain so-called *races*. For instance, Jews marry one another more often than they marry Gentiles. In so far as they do so, the Jewish minorities may be looked upon as isolates of a special type.

At the start it is interesting to try to get a grasp of the average size of isolates. By direct research upon marriages in a population we should be able to get data on which to base definite estimates, but such

Race, Reason and Rubbish

enquiries have not been carried out. The notion of an isolate has emerged only in recent years. So it has not yet become a topic for investigation of this sort. None the less, it is possible to make a rough estimate of the size of isolates. For this we can use the statistics of consanguineous marriages. If marriage between brothers and sisters went on, the probability of a brother-sister marriage in a small community would be greater than it is in a large one. It must also be greater if families are large than if the average family is a small one. The same argument applies to the sort of consanguineous unions which actually occur in human communities. In countries where they are allowed, marriages between uncle and niece, or aunt and nephew, should take place less often than mere chance dictates. There are two reasons for this. One is because they involve an age difference which discourages marriage. The other is because such unions are regarded with some misgiving as improper, if not actually immoral. Where marriage between cousins is concerned, we may possibly suspect that the situation is reversed. Even cousin marriages are sometimes considered less correct than ordinary ones; but we can hardly speak of any general opposition to them in Protestant countries. At the same time, family relationship itself involves more opportunities of social contact. To this extent we might expect that cousin marriages are somewhat more common than if mating occurred at random.

Race, Reason and Rubbish

If we calculate the size of an isolate on the assumption that marriage between uncle and niece, or aunt and nephew, is a random occurrence, we arrive at a figure too high. If, on the other hand, we rely on the occurrence of cousin marriages, we ought to get a figure rather too low, but in all probability nearly a correct one. If the calculation is made on the basis of cousin marriage frequency, such as occurs in Western European countries, we arrive at a mean figure of five hundred individuals. On the average, this means that a person has about five hundred other ones to choose from. If one refuses, only four hundred and ninety-nine remain. Needless to say, conditions are very variable. One person moves in a restricted circle, another in a large one. The figure cited is given merely as a coarsely approximate symbol of what the size of a typical isolate is like. So we should not exaggerate the consolation to be got from the refrain "there are other little fishes in the sea."

Our next job is to look at the significance of the frontiers between isolates from the standpoint of heredity. First of all, we shall deal with rare characters such as inherited defects. Presumably the genes for the latter arise by mutation. If such a new gene turns up in an isolate, it spreads itself within the latter, but seldom gets beyond its boundaries. If the boundary is more clear-cut, there is naturally less chance of the gene getting outside it. Thus the situation we should expect is that a gene for one defect will be confined to one isolate, a gene

Race, Reason and Rubbish

for another will be confined to a second, while a third isolate may perhaps lack any special genes for defective traits. It goes without saying that the probability of finding a gene of this kind in duplicate is greater in small than in large isolates. The gene may have bad luck and disappear, or good luck and spread to some extent. Only when the latter happens can two come together in double dosage. As far as congenital defects are concerned, dominant genes are of no interest in this connexion. They stamp themselves out (p. 149).

It is easy to see that the opportunity for a recessive defect to show up is much greater in a small isolate than in a large one. A defective gene which turns up in a cosmopolitan capital has no big chance of meeting its fellow. On the other hand, there need not be long delay before this happens if the same gene turns up in a one-horse hamlet. One or more defectives will then be born. So serious recessive defects ought to be more common in villages than in towns. That there are actually more deaf-mutes, morons, blind people, and so forth, in country districts than in towns depends partly on this circumstance, and partly on the fact that defectives are left behind, because unfitted to join in the exodus to the towns. If defective conditions are transmitted as recessives the condition last named is, of course, important only while the flight to the towns continues. In the next generation defectives are produced from healthy parents who carry the gene. Such individuals have no special difficulty in making a getaway to the city. Hence they can infect the town population with the genes they carry.

The conditions we have now examined furnish us with one explanation, if we want to know why inherited disease is found particularly in out-of-the-way country districts rather than in towns and built-up areas. Now and then it is very noticeable that degeneration is going on in some particular region where we meet a miscellany of morons or a community of congenitally defective individuals. It used to be said that such degeneration depends on inbreeding, but this way of putting the case is wrong. In such a locality individuals need not be of inferior stock or more degenerate than the personnel of other places. That one or another defective gene is distributed in a particular way does not signify that the population as a whole is degenerate. Whether the gene manifests itself among a large number of individuals depends on whether the isolate is small enough, and does not directly depend on inbreeding in the statistical sense of the term (p. 146), i.e. upon consanguineous marriage in *excess* of the dictates of pure chance. In another small community of the same size just as many consanguineous unions may occur without producing defective individuals, because the defective gene has not turned up through a chance mutation.

In past times isolates were smaller than they now are, and the frontiers of an isolate were sharper. To a large

Race, Reason and Rubbish

extent during the past century, and especially at the present time, isolate boundaries have broken down, or, if we prefer to put it that way, the isolate itself has grown. Growth of isolates can be brought about by two processes. One is mere pressure of population, i.e. the fact that the average number of girls who survive the age of reproduction may be greater than the number of women in each generation. Such growth of isolates has no practical effect from the standpoint of heredity. By growth of this kind the relative frequency of defective individuals does not diminish. This emphasises the fact that haphazard inbreeding has no effect. The other way in which isolates may grow depends on industrialisation, the spread of new communications, and migration to the towns. These conspire to break down the frontiers between neighbouring isolates, which coalesce to form a single large one. A breakdown of another sort, and one which must to some extent have led to the same results, is the growth of democratic institutions. Under a feudal system the nobility is quite a clear-cut isolate, because marriage with the unprivileged classes is taboo. Little by little these special isolate boundaries have broken down and some measure of class mobility has occurred. It is difficult to say anything definite about the importance of these changes, and we have little knowledge of the isolate boundaries which define the class limits of the social hierarchy.

Let us now try to get an estimate of the effect of the

Race, Reason and Rubbish

break-up of isolate frontiers. We have some information about consanguineous marriages, but only from certain regions. A case in point is Bavaria where first-cousin marriage has diminished from 0·7 % to 0·2 % in the past fifty years. We have already seen that such marriages indirectly give us some measure of the size of isolates. So from these figures we have good reason for believing that the average size of an isolate has increased fourfold in Germany during this period. Part of the growth depends on the growth of population as a whole. The part which depends on the breaking up of isolate boundaries, owing to exodus to the towns and the like, means that the average isolate has been trebled. In other countries the process may have been slower or faster. We may take it, that we may expect much the same displacement in most countries which have undergone similar changes.

We may now ask what effect this break-up of isolates may have had on the frequency of rare characters. If there is a gene for a rare defect in a particular isolate and mating transgresses the boundary between this isolate and a neighbouring one, it is not likely that the same inherited defect will exist in the latter. On the other hand, it is possible that there will be different inherited defects in the two. For instance, some type of idiocy may occur in one and some type of blindness in another. Since each morbid gene will thus spread over a larger isolate, the probability that either will get together in

Race, Reason and Rubbish

duplicate becomes smaller. This diminution is proportionate to the increase of the size of the isolate. If two isolates merge into one, defectives will be half as common. If three coalesce the number of defectives will be one-third as frequent as they were before. So the break-up of isolates must have brought about a great diminution of the proportion of all recessive defects in the population as a whole. Notwithstanding what used to be said about the growth of defects and the degeneration of civilised people, the fact is that inherited blindness, deaf-mutism, idiocy, or lameness, should now be rarer than they were fifty years ago.

A community is not necessarily degenerating because the percentage of defective individuals is on the increase. Rising figures may merely mean that registration or identification has improved. Only a fraction of defectives were previously caught in the net of official statistics. The data are now much more reliable; but they are still far from perfect and without doubt appreciably too low. Though Swedish statistics are renowned for being comparatively dependable, we have good reason to believe that the proportion of mental defectives is as much as 0·5 % in contradistinction to the official figure of 0·25 %. For Sweden the number of epileptics amounts to four thousand three hundred and fifty, but this figure is certainly too low. In general, people are not specially prone to give information about defects or diseases, but one circumstance is exceptional. We have to be more

Race, Reason and Rubbish

scrupulous about registration in connexion with medical inspection for military service. According to statistics obtained in this way there should be twelve thousand cases of epileptics in Sweden. These figures are given to show that there has been, and still is, plenty of room for an apparent increase in the number of defectives on account of improved registration.

In such circumstances there is no reason to believe that the conclusion to which we have come in connexion with the break-up of isolates is incorrect. It implies that so far as defects are concerned improvement is now going on, i.e. that the number of defectives is getting smaller. As a consequence of the break-up of isolates, the genes for defective conditions are not themselves disappearing; but they are spreading out in single dosage, and in this way become less dangerous to the community. So far as rare defects are concerned, sterilisation and other measures to hinder reproduction are ineffective. Prohibition of cousin marriages would have a somewhat greater effect; but this possibility is not at present feasible, and for various reasons its desirability is debatable. In the meantime a real diminution of defectives is being brought about by break-up of isolates. In so far as we are concerned with rare defects, the social instruments which make for racial betterment are the bicycle and the omnibus, the flivver and the jazz band—in short, the collective enjoyments of town life.

When we discuss problems of race biology we often

Race, Reason and Rubbish

lay emphasis on the elimination of the unfit, that is to say on sterilisation and similar measures. This is short-sighted. A community may have more or less defective individuals without much harm to it. The expense and inconvenience which they entail for the community as a whole does not amount to very much, even if we are seriously determined to provide them with a really decent existence. What does matter is the general intellectual level of the population. Within limits a community does not degenerate because the number of defectives goes up. It does degenerate if stupidity gets the upper hand and the general level of intelligence sinks. We may therefore ask what influence the break-up of isolates has on inherited traits which promote social efficiency—in other words, how the process affects ordinary characteristics which are inherited in a complicated way.

Another character may be used to illustrate the possibilities which occur. The average height of Swedes has increased 9 cm. in the course of the last century. The available data are based on universal medical inspection for military service and are therefore dependable. It is natural to assume that this increase is due to a better standard of living, and agencies of this sort may well have contributed to it, but it is also difficult to believe that we can entirely explain increase in body length by recourse to them. One reason is that the severe famine years in the 'sixties left no mark. There was no fall in

Race, Reason and Rubbish

the figures. Another is that the standard of living has improved more in recent years than during the first part of the nineteenth century. So if tallness depended on the standard of life, we might expect that the improvement happened more quickly in recent times than in former times. In reality it has been uniform.

Space does not allow us to go more deeply into the problem of the mechanism of growth, but it seems clear that this increase in adult stature does not wholly depend, and perhaps does not depend at all, on a rise in the standard of living. If so, it must be due to changes in the gene equipment of the population, and it is plausible to believe that the break-up of isolates has played a powerful part in producing it. Stature depends on a number of genes which reinforce one another. It is reasonable to assume that different genes occur with different frequencies in different isolates. Through the break-up of isolates these genes have been spread out in single dosage. If tallness depends on dominant genes, the break-up of isolates must therefore bring about increase of stature. At present we cannot venture a dogmatic statement about this, but it is a theoretical possibility which seems to be likely at the moment.

It is also possible that intelligence behaves in much the same way as stature. With reference to stature, there is a spread of variation from isolated very tall individuals through people of moderate height to isolated individuals who are very short. Besides these

Race, Reason and Rubbish

there are excessively short (or tall) individuals owing to the existence of exceptional genes, dwarfs or giants produced by special very rare genes, and individuals who are stunted by morbid conditions. Apparently intelligence and other aspects of social behaviour are distributed in much the same way. The range of variation extends from genius through persons of moderate ability to the really stupid. Beyond the latter are mental defectives, who have abnormal characteristics for various reasons, e.g. because of special genes, because of infectious diseases of childhood, because of syphilis, or because of cerebral bleeding at birth.

The break-up of isolates means increased frequency of heterozygotes, i.e. of dominant types. If the break-up of isolates affects the general intellectual level as we have supposed that it affects bodily stature, it would seem that the number of stupid people has gone down while the number of extremely gifted ones has gone up. For this to happen, mental characteristics must in general depend on dominant genes. Is it really possible to believe that such a change has come about? To reply to such a question we have to have more knowledge of the intimate nature of the hereditary mechanism which lies at the roots of intelligence and similar characteristics.

Researches carried out so far hardly give us any foundation for an answer to this fundamental question. We must be content with stating a possibility. It may well be that such characteristics depend on a mixture of

Race, Reason and Rubbish

dominant and recessive genes, so that, after all, the mechanism is not so significant, but it is not likely that there is a complete balance. If dominant genes are mainly responsible for advantageous traits we have prospect of improvement before us. If less advantageous genes are dominant, we are faced with decline; and we have obviously no grounds for excluding the possibility of a serious fall of the general intellectual level. Either way, the processes discussed cannot be easily restrained. What has happened or is now happening cannot be undone.

CHAPTER XI

In what has gone before racial groups have been mentioned and race-crossing has been referred to merely as a special case of the *isolate effect*. Unless a racial group constitutes an isolate, or was formerly one and is now in process of breaking up, it has obviously no special interest for us. If such a group is distributed throughout a population, we may possibly talk of the latter as a mixed race, but there is then no point in talking about the races themselves. What may be said of isolates in general thus applies to races in particular, but the issue has other aspects, because it has now entered the arena of politics. People discuss the race problem in general and most often without the least elementary acquaintance with principles of inheritance and the processes which determine changes in the hereditary make-up of human populations.

The conception of race was originally used in connexion with plants and, as such, it is not acclimatised to the intellectual atmosphere we usually inhale in contemporary discussion about race problems. When we now discuss crop improvement we have to talk about definite characters or genes and define precisely what we mean by them. In the practice of animal breeding, earlier notions about race continue to exert some influence; but, even so, people are beginning to discuss

Race, Reason and Rubbish

characters and genes intelligently. When human beings are the topic this is by no means true. Even in scientific circles, we sometimes use the word race without trying to be clear and definite about what we really mean by it. An illuminating comment on the situation is the fact that it occurred in the title of only five out of three hundred and ninety-six papers read before the International Congress of Genetics held at Edinburgh in 1939.

In the language of everyday life we cannot always give clear definition to every word or concept we make use of, and it is not necessary to do so, because there is often no danger of misunderstanding. In scientific research, on the other hand, it is essential to work with clear-cut definitions. Every branch of science therefore evolves its own terminology, a collection of symbols each with a definite meaning of its own. One of the reasons why it is difficult to popularise scientific discovery is that it is almost impossible at times to translate into everyday speech a statement set out in correct scientific terms. As happens in this book, we have sometimes to get to grips with unfamiliar words which may have a foreign and complicated sound, but are none the less words which make it possible to distinguish essential principles or facts.

So if we are to talk about race questions, it is quite essential to start by discussing what we mean by a race. We talk of two races when there is a tangible difference between one group and another. Such a difference must

Race, Reason and Rubbish

be determined by heredity. We have thus to ask ourselves to what extent inherited differences between individuals must exist before we regard them as members of different races.

When we try to classify systematically all sorts of organisms which exist on our earth, we naturally begin with the most fundamental differences. For instance, we divide them into animals and plants. On account of recognisable differences we eventually put them in different families, genera, and species.

Species are the smallest groups which have sharp boundaries. Within species there may be smaller groups which have not such clear-cut limits. They may be called *races* or geographical varieties when they occur in nature, and breeds, strains, or pure lines, when they are the product of human interference. A straightforward definition of race therefore involves the possibility of distinguishing between races and species, hence also of defining the latter.

Unfortunately there is no universal agreement about a definition of species except in general terms as stated above. Some biologists, especially botanists, prefer to restrict it to the smallest groups of which the constituent members will not breed effectively with those of another. If members of a species defined in this way do mate with members of another, and so produce offspring, the latter are sterile like the mule. They must therefore preserve hereditary differences which distinguish them, even if they share the same habitat. Though this criterion has a precise meaning from the standpoint of the biologist who is interested in the bearing of heredity on evolution, it has never been adopted by all people occupied in the practical task of cataloguing the diversity of living creatures. One reason for this is the large number of types—especially plants—which reproduce by self-fertilisation or virgin birth. Where this occurs, any detectable mutation would provide a sufficient basis for making a separate species. Another difficulty arises from the fact that opportunities of hybridisation, such as those which search for new varieties in connexion with the practice of horticulture and crop improvement entail, are not available in zoological museums. The zoologist has to deal largely with dead material and cannot get seeds sent by post.

Workers in museums therefore prefer to adopt a purely practical standard, and the common custom is to separate species when such groups maintain their distinctive characters without interbreeding and producing intermediate types. This does not necessarily entail the impossibility of producing fertile hybrids under experimental conditions. The natural limits may be set by geographical barriers of water or mountain, by ecological habitat, and by habits or season of reproduction. If the topographical or ecological barriers which separate two groups do not completely exclude mutual accessibility they may interbreed and produce intermediate types in

Race, Reason and Rubbish

regions where their respective territories overlap. When this is so, they are usually denied specific rank, and are referred to as races of the same species. It follows that the partition between race and species is often thin. Species and races are subject to revision as we get to know more about the distribution of animals and plants. Even when this qualification is made, it must also be admitted that the practice of zoologists and botanists is not completely uniform.

The fact that the territories of two groups do not overlap, generally means that they have been separated for a longer period than two groups which hybridise in nature. So more time has been available for the accumulation of different mutant genes. This also means that the number of hereditary characters which distinguish two closely related species is generally greater than the number that distinguish two races within either of them, but it is not a general rule that the magnitude of the gap between the characteristics which distinguish species is always greater than the magnitude of the gap between characteristics which distinguish races. Much depends on how far the end in view compels the systematist to pursue his observations. If it is very important to do so, as for instance because one fly may carry a disease and another may be immune, species may be delimited by very minute differences. Similar remarks apply to races. What is most important from the standpoint of subsequent discussion is that biologists do not speak of two

Race, Reason and Rubbish

races in nature unless each group actually has some territory where interbreeding does not occur appreciably. If two types of animals or plants distinguished by hereditary characters inhabit the same region, they are not assigned to different races, unless there are neighbouring regions where each type is more or less completely localised. Thus a race like a species must be an isolate or group of isolates.

So far as we know all recognisably different types of human beings in different parts of the world can interbreed and have actually done so in some times or places. So the differences which distinguish them do not justify us in regarding them as different species. Within the single human species there are some more or less clearly defined local varieties visibly recognisable by their appearance. Between negroes and Chinese are differences of skin colour and other dissimilarities which distinguish any two individuals belonging to these two races. They are repeated regularly in the offspring of pairs which belong to them; and there are parts of Asia and Africa where the population is made up exclusively of one or the other type. Between Asiatic Mongols and the white people of Europe there are constant differences of skin colour and the set of the eyes. In general Mongols are shorter than white people, but this is not a rule which admits no exceptions. A collection of other characters, of which some are clear-cut and others are a matter of degree, also distinguish them.

Race, Reason and Rubbish

All such differences involve externals. People often assume that there are other differences which involve the way which organs carry out their tasks. Perhaps we should not go so far as to say that the kidneys of a Mongol do not operate in the same way as those of a white person, or that liver and heart are essentially different from a structural point of view; but some people are quite ready to assert that the brain of a Mongol does not work in the same way as the brain of a white man. Somehow the soul is in a class apart. In some way it is supposed to be coloured by the appearance of an individual. In reality Mongols do not necessarily use a different sort of logic. They entertain affection for their children or near relatives as we do. Like ourselves they wish to get on in the world and to have an agreeable existence. They can show strong acquisitiveness and great greediness, as they can also be helpful, friendly, and self-sacrificing. They have social impulses like our own and a limited capacity for rational thought as we have. But in spite of these similarities we still suspect the existence of essential differences. The curious thing about this is that if we read the literature of race "psychology" we find that there is only one thing people are agreed about. They all say that there really are differences, but when any particular race is concerned, they are not able to reach agreement about what psychical peculiarity is found in one race and absent in another.

Race, Reason and Rubbish

The real truth is that, when we discuss race we backslide to antiquated notions about heredity. Before Mendel's time people believed that the hereditary make-up of an individual is a homogeneous substance which can be mixed and diluted. Mendel showed that it is really a collection of separate particles which may not be the same in different individuals. Two individuals may have many genes in common but also differ with respect to others. The race concept is an old one and established itself a long time before Mendelism laid the foundations of the science of heredity. As we might expect, some people have not kept in step with this development. They imagine, more or less definitely, that the essential stuffing of two races may be as different as copper and quicksilver. It is not merely a matter of one or more characteristics which depend on one or more particular genes. We are back to what Aristotle called the *essence* of material bodies. As copper and quicksilver have molecules which are not built up in the same way, two races have different "blood." As copper can be dissolved in quicksilver to produce an amalgam, pure races can blend to form mixed ones.

People talk of pure races and mixed races as if they were always discussing clear-cut differences of one and the same magnitude. We too easily forget that the differences between different races are not of the same order. We can define race differences clearly, and only, with respect to trivial characteristics which depend upon

Race, Reason and Rubbish

external appearance, and so forth. Though we ought never to assert that differences occur when we really know nothing about them, the fact that we do so would not matter so much, if the issue at stake were merely theoretical. Naturally, we are entitled to discuss a possibility, provided that we do not discuss it as if it were a matter of fact. Perhaps also it would be more excusable if discussion of this sort were confined to major groups, but such reasoning may obviously be more dangerous when exaggerated statements are made about divisions within the chief races of mankind. Negroes and white are distinguished from one another more noticeably than different European peoples. As far as the inhabitants of Europe are concerned, it is self-evident that similarities are relatively great.

To be sure, there are certain differences even in Europe. For instance, South Europeans are more often sallow, brown-eyed, and short of stature than North Europeans, but when we try to get clear about the occurrence of such differences between different people, we find that it is impossible to make hard and fast rules. Thus an Italian may be blued-eyed, pale, and tall, while a person born in Sweden and descended from a long line of Swedish ancestors may be dark, brown-eyed, and short of stature. The way to set about discussing these things is to try to find out how differences manifest themselves among different peoples, and rest content with recording them in percentages of persons who are

Race, Reason and Rubbish

brown or blue-eyed, short or tall, etc. We should then have definite information about how characters are distributed. Perhaps we should not succumb to the temptation of assuming psychical dissimilarities supposedly connected with different kinds of external traits which occur in different communities.

Instead of this we now have theories about the origins of pure races which have been blended by migration of tribes, nations, or individuals. This is supposed to explain why no community nowadays corresponds to a pure race. In reality, this involves an assumption which lacks any pretence of certainty. There is no reason whatever for presuming that there ever was in Scandinavia a pure Nordic race subsequently contaminated in this way. According to an examination of army recruits undertaken in the years 1897–98 to analyse the racial make-up of the Swedish people, only 10% of them were classified as examples of the pure Nordic type.

The fact is that we now talk of race in two different senses, at times pseudo-historical and at times biological. We adopt the pseudo-historical point of view when we assume a past existence of pure races which have intermingled so that they are no longer really pure to-day. We speak of races in the biological sense of the term in so far as hereditable differences distinguish two groups. To make the distinction clear we may suppose that only black Andalusian fowls are kept in North Germany, that only white ones are kept in the south,

Race, Reason and Rubbish

and that they interbreed in the middle, so that we there find some blue Andalusians. The latter do not breed true to type, but yield blue, white, or black chicks. From a strictly biological point of view we have three groups each with its own inherited make-up and each with an ill-defined frontier. If we call the group in the middle of Germany a *mixed* race, in contradistinction to the two pure races of the north and south, we make an historical judgement about the way in which it came into being. That is to say, we imply that the various types of individuals which make up the middle group arose from the mingling of two groups which were originally distinct. While this may be, it is not necessarily the correct explanation of the facts. In nature the reverse is often true. That is to say, a mixed group may be the common parent of the pure types concentrated by natural selection in opposite regions.

In any case, our knowledge of how similar combinations turn up among hybrids shows that individuals who resemble one or the other pure type are not necessarily related more closely to similar than to unlike individuals of the pure stocks from which a hybrid population has arisen. Fowls with *walnut* combs in the first hybrid generation of a cross between purebred fowls with *rose* and *pea* combs (p. 77) are more closely related to either of their pure-bred parents than are homozygous fowls which turn up with rose or pea combs in a subsequent generation. It goes without

Race, Reason and Rubbish

saying that such external similarities are no guarantee that a given individual in a hybrid population will have inherited desirable but less tangible traits of a pure-bred ancestor with the same external appearance as itself.

In this connexion we should remember that we may talk of pure races among animals when there is at least one inherited trait to distinguish them. It seems that when we talk about two races of human beings, we do not mean a single definite difference of this kind. If so, we might as well assign every individual to a special race, because no two individuals are exactly alike. If we want to make a complete analysis of two populations from the standpoint of heredity, we have to distinguish situations in which particular genes found in one group are lacking in a second from situations in which the differences are less fundamental and merely involve the fact that some genes are more common in one group than another. We should then assign definite numbers to homozygotes or heterozygotes with respect to each pair of genes, and at the end of our task we should probably find reason for concluding that differences between such groups are trivial in comparison with differences between individuals.

This much is certain. We never meet with two individuals who are exactly alike. With respect to important characteristics, we often find that two individuals belonging to different groups agree with one another more closely than with particular individuals

Race, Reason and Rubbish

belonging to their own group. In so far as this is so, we ought to judge every individual according to his or her own hereditary make-up. In so far as we can do so, there is no point in stating whether he or she belongs to one race or another. At present our information about such matters is defective. We are not able to report exhaustively on the make-up of individuals or of groups. We have to rely on average differences of greater or less degree. Thus the notion of race, though valid as a descriptive label, is an unsatisfactory makeshift when we look at the matter scientifically.

We have now to ask how matters stand when we try to set definite limits to human races. It has been the custom to start by imposing special demands on what we mean by a typical representative. Thus the Nordic type is represented by people with blue eyes, long skulls, comparatively light hair, tall stature, and so forth. To find out the extent to which persons of pure Nordic race occur in a community, the procedure has been as follows. Persons with brown eyes are first rejected. From the remainder short individuals are lopped off. From those left people with short skulls are excluded. So the process continues, and it is obvious that if we use many characteristics, we are left with very few individuals. What is still more important is that the limits we set are quite arbitrary and that there is no unanimity about where to set them. There is every gradation between blue eyes and brown ones.

Race, Reason and Rubbish

What trace of brown we can tolerate in a blue or gray pupil without assigning a person to a mixed race is still a matter of opinion. What stature we must reach to belong to the Nordic race is undecided. As we all know, brothers and sisters can be very different. It may happen that one is short and has brown eyes, while the other is tall and pale. It is plain nonsense to state that a brother and sister belong to different pure races, as, for instance, that one is Nordic and the other is Alpine.

In contemporary discussion about race we therefore come up against two problems. One is to discover to what extent we really mean anything at all when we talk about individuals belonging to different race types. The other is to discover the significance of group differences. It ought to be clear at the outset that we cannot speak with any assurance about whether a person belongs to one race or another when we are discussing Europeans, though there is no difficulty about deciding whether a person is a negro or a white.

To decide whether there is any meaning in talk about race types within a community, we have to ask if there is any connexion between inherited characters. To clear the ground for action, let us suppose that negroes and white people live together in the same country and that each group is at first an *isolate*, so that interbreeding does not occur. If we then made a study of the complete population, we should be able to recognise a constant

Race, Reason and Rubbish

association between certain characters. For example, a person with curly hair would also have black skin and would be a negro. Let us now imagine that the boundaries of the isolate break down, and that miscigenation occurs between the negroes and the whites. The connexion between the characters is consequently weakened. After a number of generations we could no longer argue the presence of one character from the fact that a particular individual may have another. In spite of the fact that an individual had straight hair he might have a black skin. Contrariwise he might have a light skin combined with curly hair.

The reader may here ask how quickly the connexion is dissolved after the complete breakdown of the isolate boundary. To answer this we need some way of measuring the association between two characteristics. An index often used by statisticians is called the *correlation coefficient*. The value of this is 1·0 when the association of two characteristics or measurements is perfect, and zero when association is quite haphazard. Mathematical analysis shows that the correlation coefficient for two genes is diminished 50 % in each generation, when free interbreeding is allowed to take place. This means that the connexion between them is too small to bother about after the lapse of a few generations. So even if there really are significant inherited psychical differences between whites and negroes, any connexion between psychical traits and outward appearance would disappear

Race, Reason and Rubbish

after a few generations of miscigenation; and if it were true that negroes have black souls instead of pale ones, we could no longer come to rational conclusions about the connexion of soulfulness and skin colours. After a comparatively short time an individual with particular external traits would be equally likely to have a black soul or a white one.

If we are in a position to state that there is no known connexion between physical characteristics, it is meaningless to talk about different race types within a community. If we exclude the region where Finns and Lapps are comparatively numerous, researches carried out on the Swedish population to discover such a connexion between physical characters have shown that no such connexion exists. It may be argued that there is a Jewish section which obviously constitutes a group apart, but in Sweden Jews are not numerous. In one year group of recruits, i.e. forty-five thousand twenty-one-year-old males, there were only about fifty Jews, and this number is too small to justify a statement about the connexion between inherited characteristics.

When we turn to other parts of Europe, the scope of research into physical characters as a basis of racial classification has been meagre. During the year 1886, R. Virchow carried out an enquiry embracing 6,758,827 school children in Germany, and subsequent researches have been undertaken with relatively scanty material. Topinard (1889) made a study of two hundred thousand

Race, Reason and Rubbish

individuals in France. In Italy Livi made a similar enquiry during the year 1896. This dealt with three hundred thousand individuals. Nothing more comprehensive has been attempted in England. All such enquiries have been carried out with defective technique, and the data have never been worked over in a satisfactory way. In many ways they are now obsolete. All the more recent investigations are based on small samples, and, even so, are deficient from a scientific point of view.

We really know little about body-build, eye colour, and such characters among the peoples of Europe, and the gap in our knowledge has unhappily left a free field for wild speculation. If we exclude a few comparatively small groups such as Jews and gypsies, there is scarcely any reason to believe that we should find much evidence of definite racial groups with single countries from established facts about the wanderings of peoples in ancient times, or about the exodus from country to town, and other effects of modern communications. Even if such groups exist to an appreciable extent within a single nation, it does not follow that there are representative differences between one nation and another. In reality it is not very likely that very significant differences of this sort do exist.

The assumption that pure races such as the Nordic or Alpine exist, or have ever existed, is purely a hypothesis which has little scientific basis. The extent of migration or interbreeding across the frontiers of

Race, Reason and Rubbish

nations has been so great that there is scarcely any reason to imagine that essential differences occur. There are obviously some differences between North and South Europeans, but there is no more room for large ones between the Germans and the French, or between the Germans and the English, than between North Germans and South Germans, or between Frenchmen from the North and South of France. Obviously, also, it is not very reasonable to believe that citizens of the United States constitute a peculiar racial mixture. The American people has arisen by a blending of immigrants from all European countries, and there is no definite basis for saying that the American is essentially different from the European melting-pot.

At the beginning of this chapter it has been emphasised that there is no point in talking about a race except in so far as such a group is an isolate or has recently been one. Tall people who exceed a certain body length have an inherited character which distinguishes them from the section of the population below this arbitrary limit, but no one has yet hit on the idea that they make up a race of their own. This is because they do not form an isolate. They mix and mate with the majority of short people. The notion of an isolate has been put forward quite recently, so ethnographers have not yet taken the precaution of finding where the boundaries of isolates lie before reconstructing races symbolised by race types based on considerations in which aesthetic pre-

Race, Reason and Rubbish

judices play a large part. We may seek in vain for a really ugly individual among the illustrations of countless works where the Nordic race is discussed. Since we now know that isolate boundaries are always breaking down by migration of peoples, by gradual displacement of one kind or another, as also by the movement to the towns, and other consequences of modern transport in more recent times, we have little reason to imagine the existence of isolate boundaries which circumscribe race groups notably distinguished from one another among civilised peoples.

Even the differences between nations seem to be trivial ones. The hatred between them, now driving Europe to destruction, cannot be justified by appeal to inherited differences established by scientific research. Between nations there are differences of tradition and culture, speech, habits of life, behaviour, government; but such differences do not depend on different racial characters. Some people have maintained that they could be produced in no other way. We have been told that they must be the expression of how human nature itself differs. Needless to say, this assertion is not correct. Different economic and social circumstances, different systems of schooling and the like, can assuredly produce fundamental differences which have no hereditary basis.

The peoples of the world have not had equal access to its national resources. Chance circumstances have

Race, Reason and Rubbish

deflected social evolution in one direction or another; but the social organisation of western countries has followed much the same track from feudalism through absolute monarchies which have more or less completely broken down to democratic states in which class conflicts play an important role. We have passed from peasant production to industrialism and the process has been swifter in some countries than in others. Differences of culture and social life exist, and we may have occasion to emphasise them, but there is no proof that they are connected with inherited differences between the peoples concerned. When we get clear about the fact that such differences, whether small or great, depend mainly on external circumstances, it will be difficult to work up real hatred against other people. Scientific enquiries which clarify the issues and the popularisation of such research should therefore mellow opposition, and force people to realise that the circumstance of getting born on different sides of a frontier is no reason for hatred.

An opinion specially sponsored in Germany, though curiously enough one which has failed to arouse enthusiasm in Scandinavian countries, is that the Nordic race overtowers all others. This type is specially common in Scandinavia, where there are comparatively many people with blue or gray eyes, fair hair, long skull and face, tall stature, and a comparatively slim build (p. 201), but when racialists of other countries visit Scandinavia they are often disappointed to see that middle-aged

Race, Reason and Rubbish

Swedes, Danes, or Norwegians, are somewhat inclined to corpulence. The ideal Nordic type of their dreams is not so common as we are apt to think. According to the estimates of Fürst and Retzius (p. 202), even Swedish recruits who come up to standard are only 10 % of the total. Though exact figures are lacking, the proportion in Germany is certainly less, and in South Germany it is still lower than in the North.

Nobody knows where its boundaries lie. So it may happen that the Nordic standard is defective, and that the real Nordic type is even more outstanding than it is supposed to be. Those who think so may believe that a relatively small infection may be enough to ensure the moral, mental, and physical superiority of a nation. It requires little sophistication to realise that sufficient reason for the assumption of such superiority is lacking. To be sure, the Swedish people is not specially below par. When those of us who are Swedes peep into our past we can say that we have sometimes acquitted ourselves well, and sometimes distinguished ourselves less conspicuously. We have won our wars and lost them like other nations. A number of outstanding scientists and artists have been Swedes, but many quite mediocre people and even some imbeciles have also been Swedes. We have insufficient reason for asserting that the average level has been higher than in other lands. For the time being, Scandinavia enjoys a great measure of respect abroad. We have had a long

Race, Reason and Rubbish

spell of peace and we have been fortunate in other ways. So in many respects social conditions of Scandinavia are better than those in other lands; but our raptures over such differences should not drive us to the nationalistic creed that our peoples have a special genius and a special superiority.

If we look at other lands in which a greater or smaller proportion of people are supposed to belong to the Nordic race, the state of affairs is very variable. Both Britain and Germany have played an outstanding role at one time or another. The French were once top nation in Europe. Italians may look back with pride to centuries when they had an effective influence and were leaders of European culture. All peoples have proud memories and bad periods they prefer not to think about. In the few generations which make up the recorded history of Europe we can scarcely say that any nation has significantly shown itself to be the superior of all others. Some people have tried to support Nordic superiority by the assertion that genius goes with good looks of the Nordic type. The truth is that genius wears a motley garb, and no one has succeeded in making a case for the conclusion that genius is more likely to be blue-eyed and tall than brown-eyed and short, or to be domiciled in a long skull rather than a broad one.

Among the more specious proofs once brought forward on behalf of Nordic superiority is an enquiry carried out in Bavaria. It led to the result that town

Race, Reason and Rubbish

dwellers were somewhat more Nordic than villagers. It was taken for granted that the townsfolk were more talented than people who lived on the land, and that the town population was selected from the more gifted individuals of the surrounding country districts. The assumption that the average townsman is more gifted than the average countryman may be right, but it is not self-evident. The assumption that the immigrant from the countryside is largely of Nordic type is also open to question. In reality the town population is not merely recruited from the surrounding country, but is far more mixed than the population of the latter. In these South German towns there are presumably more North Germans than in the country districts around them. Since blue-eyed blonds are commoner among North Germans than among South Germans, this would itself lead to some difference between the population of town and country.

In other investigations involving comparisons of town and surrounding country districts opposite results have been obtained. It is a suggestive fact that, according to the enquiry of Livi (p. 209), students in North Italy are more often brown-eyed and short than the rest of the population. The prosperous classes are a more well-defined isolate than the lower classes. The families from which students come are somewhat mixed with South Italians, who are shorter on the whole and more often have brown eyes than people in North Italy. So enquiries

Race, Reason and Rubbish

of this type have by no means strengthened the assumption that the Nordic race is superior.

Neither with reference to the nations of Europe nor to special groups within each nation have we really any reason to believe that there are qualitative race differences. Nor have we reason to believe that any particular people is appreciably worse than any other. Some confidence in his own capability and some self-respect is necessary as well as advantageous to the individual. It is difficult for people to believe that they are inferior, but it is positively dangerous when individual self-respect exceeds the bounds of what is reasonable. Up to a point self-esteem is advantageous and desirable for groups and peoples as for individuals; but self-esteem can be dangerous when it is excessive and overrides the sentiment that we are all human beings who might be helping to shape a better future.

An issue which is not of great significance in Europe but is very important elsewhere, specially in the United States and in South Africa, is the endowments of the negro. It is generally held that the negro race ranks low in the scale of human values, and if this is merely another way of stating observed facts without any definite implication concerning the role of nature and nurture, the matter is clear enough. In Africa negroid peoples still live in many regions untouched by what we ourselves call civilisation, and though such a state was at one time thought to imply something akin to

Race, Reason and Rubbish

the innocence of childhood, we do not usually idealise it in such terms to-day. In places where they have come in contact with western civilisation their initiation has not happened under specially encouraging circumstances. In America they were introduced as slaves, and elsewhere they have been exploited by imperialistic nations who adopt towards them the tone of forbearing intolerance which grown-ups show to children. From the start they have therefore been stamped with the hall-mark of inferiority. Because of this attitude on the part of the white man, they have been excluded from opportunity and have been victims of a sort of social boycott. The negro comes into the world with a skin which darkens quickly after birth, and with a brain which soon blackens before the realisation that he must abandon all hope. We do not know what might happen if negroes were treated as social equals with access to the same privileges as white men.

To be sure, so-called intelligence tests have been carried out on whites and negroes, but such tests have not given unequivocal results. In some enquiries striking differences have been recorded, in others nothing of the sort. A peculiar result was obtained when negroes and white recruits for military service in America were subjected to such tests during the last world war. Negroes were significantly inferior to whites, as were soldiers in general to their officers, but negroes from the Northern States scored higher than whites from the

Race, Reason and Rubbish

South. This enquiry has been sharply criticised, and with some justice. In any case investigations carried out so far give us scarcely any justification for making a decisive statement about the extent to which differences of social efficiency between negroes and white men depend on heredity or environment. The situation is clear only in so far as some negroes clearly have an inferior intellectual equipment to some whites, and some whites clearly have an inferior intellectual equipment to that of some negroes. We need more comprehensive research work before we can say more than this.

Meanwhile we have a reminder of the influence which external circumstances may exert on peoples of European stock in the existence of communities in some parts of South Africa and the United States, where large numbers of poor whites live in degraded conditions. There at least we find grounds for believing that particular social conditions can mould a people. For the rest, the negro problem covers so many facets of social and economic history that a thorough-going treatment of all the questions involved would call for a book by itself.

CHAPTER XII

THE Jewish question has many sides—religious, economic, social, and racial. We cannot here discuss it in a comprehensive way. What follows deals briefly and chiefly with its racial aspect and incidentally with related questions. To start with we have to recall the fact that the Jewish question was originally a religious one. It was a quarrel between Jews and Christians. In the time of Christ the prevailing opinion among Jews was that they had broken away from Jehovah and were being punished on that account. The Christians believed that they had secured the forgiveness of God through the propitiatory death of Jesus. Their doctrine was that Jesus was the only begotten son of God and had given himself for the human race. He submitted to suffering beyond endurance, and through his sufferings cancelled a debt which we ourselves could never pay. Hence men and women could feel free of guilt apart from individual sins committed before repenting and seeking forgiveness. Jews believed that the debt had never been paid and Jesus was not Messiah.

Many Jews still believe that Jehovah is displeased with them for misdeeds committed by their ancestors. If one accepts the Jewish view of the deity this is the only possible opinion. The explanation of the fact that Jehovah permits the misfortunes of his chosen people

Race, Reason and Rubbish

can only be that he is angry and that his wrath must be appeased. Perhaps the Jew who can believe this finds some comfort in his creed. He need not believe that human beings are debased or cruel and that people must still, as formerly, cause one another measureless distress to gain trifling advantages, or no personal advantage at all. Faith in an angry deity obviates shame and in some measure ennobles suffering. The Jew who is persecuted sees his misfortunes as a necessary sacrifice which brings his fellows nearer to the Messianic age, when there will be no more suffering.

Before the nineteenth century, and from early times, Jew baiting had always had a religious *motif*. The Jews were the personification of Judas, the betrayer. They were the people who had rejected saving grace. They were not entitled to rights because they had set themselves unforgivably against the Son of God. When times were bad, and specially after war or during a famine, hatred of the Jews has flared up and has led to pogroms. This has never completely died down. At some times and in some countries Jews have been treated with tolerance and to some extent have enjoyed civic privileges, but they have been forbidden to possess land and to hold public appointments in all Christian countries. Their opportunities of livelihood have been limited, often very much so, and they have always been beset by antipathy.

Latterly anti-semitism has flared up again and with

Race, Reason and Rubbish

greater virulence than ever, but the new anti-semitism has a rationale quite different from that of its prototype. Though the traditional religious point of view still lurks in the background, it is no longer a matter of faith. The Jew who has taken the precaution to be baptised or even an individual more or less remotely descended from Jewish ancestors may be looked upon as an inferior. The new fashion is that the Jews are a *race*. This race possesses regrettable characteristics and must therefore be eliminated. The problem is envisaged from a national view-point. It is said that people of foreign blood should not be allowed to live within the frontiers of a country. Since they have no land of their own, the Jews would not be able to exist anywhere upon the earth, if such a view became universal.

The racial point of view first began to be influential during the nineteenth century. Although it had long been commonplace that there are differences between white men, Redskins, Chinamen, and Negroes, and that we can therefore classify human beings in races, racialist views had not previously obtruded into public discussion—at least in relation to Europe. It is undeniably odd that the prevailing racial superstitions of Germany were moulded by a Frenchman and an Englishman. The first person who tried to evolve a thorough-going doctrine that race similarities have a significant bearing on the qualities and destinies of nations was Gobineau, at one time French ambassador at Stockholm. Among

Race, Reason and Rubbish

the earlier authors Houston Chamberlain, an Englishman who lived in Germany, is also noteworthy. Both of these men put forward the opinion that the Nordic race is superior to all others. Neither of them were scientific men. So it is easily understood that they were not specially critical about what they asserted. Scientific discussion, from which we are in a better position to distinguish what we know from what we do not know, came later. Thus research which gives us a critical and reliable summing-up of the scientific position must still, and even in the immediate future, expect to meet violent opposition in some quarters.

Of course, there is no point in discussing the Jewish question from a scientific point of view with individuals who embrace extreme anti-semitic doctrines. There are many others who do not go so far. They tell themselves with some hesitation that anti-semitism must have a basis. On the whole, therefore, Jews must be rather bad people, though there may be exceptions. Many of them also think that it is justifiable to treat morally reprehensible people with a hard hand so long as such treatment does not extend to unnecessary cruelty. At an early age nearly everyone is infected with a more or less pronounced anti-semitism. We get to hear that the Jews are unscrupulous in business, that they are morally below par, that they are not loyal to the peoples among whom they live, that they are not good patriots, that they do not take to agriculture or work with their

Race, Reason and Rubbish

hands, that they speak with a foreign accent—and much else. Besides all this, they possess hooked noses, are obese, have flat feet like policemen, and, taken all in all, are not easy on the eyes. Worst of all, they wield a disproportionately powerful influence. They are too apt to belong to the prosperous classes, to be wealthy, skilled in medicine, or trained for the law.

Thus we accuse the Jews of a bewildering variety of less attractive characteristics. The only thing we do not charge them with is stupidity. We make out that they are not well thought of, but in spite of this they manage to be popular as doctors or lawyers and, as such, have unnecessarily large practices. The Jews are unreliable and too clever by half; but we do business with them and buy their goods inordinately. Business men declare that Jews are mainly business men, as if such work were specially deplorable. Doctors point out that too many Jews are doctors, as if there were something perverse in getting paid for healing the sick. Presumably, there are also lawyers, who consider that even the prosecution of their own profession can be unworthy when it is the Jews who are carrying it on. We do not maintain that the Jews are bad business men, bad doctors, or bad advocates. Quite the contrary.

As a rule we respect success. We consider that a man who gets on in the world must be competent. None of us deny the competence of the Jews, but we make out that they owe their competence to bad morals. If we

Race, Reason and Rubbish

look into their history and see for ourselves how cruelly and brutally Christians have treated them at all times, or if we merely read in ordinary newspapers about how the Jews are handled in civilised lands to-day, we may well wonder if the ethical level of Christian countries permits us to look down on them from a moral attitude. Such sidelights on the problem as we thus get scarcely lead us to the conclusion that Jews are much worse than Christians.

The extent to which genuine racial differences distinguish Jews from other people still remains a problem. In appearance it is obvious that they are often different from people among whom they live. So one question often discussed is whether the Jews should be considered a pure race. The opinions of ethnographers about Jews may be diverse; but there can be no doubt about the fact that they do not fulfil the conditions we are accustomed to expect when we talk about a pure race. There are brown-eyed and blue-eyed Jews, Jews with dark hair, Jews with red hair, and Jews with light hair. As far as looks are concerned Jews have a wide range of variation—like other people. This does not exclude the possibility that there are average differences, and there are many people who state that they can tell whether a person is or is not a Jew at first sight.

In Scandinavia, where the differences are more striking than in Germany and Italy, the author has tried to test how far such people can really distinguish Jews from

Race, Reason and Rubbish

Swedes. In some cases an individual may make a correct diagnosis, but in many others judgement is false. From time to time the author has invited race experts who make out that they can recognise a Jew from a photograph, to pick the Jews in his treatise on twins containing 180 pictures of twin pairs. When the twins were investigated all that could be got together were photographed. So the material is unbiased. But the author himself has never met anyone who could guess correctly in more than a few cases.

Some anthropological researches show that Jews are more or less reminiscent of the peoples among whom they live. For instance, there are relatively more brown-eyed short Jews, if they live among people who are predominantly dark and short. So far back as we can go the Jew has always been a variable type, and Jews have always mixed to some extent with people around them. From this point of view we must make a distinction between Eastern and Western Jews. The Western branch came to Europe *via* North Africa and Spain, the Eastern across the Balkans or Poland. Between the two groups there are some average differences. Indeed, we may go so far as to say that Jews are a divergent type in comparison with Scandinavians, but the magnitude of the difference is of the same order as that which separates Swedes from Spaniards or Danes from Italians. Nowadays there will surely be few so bold as to assert that an average divergence of external

Race, Reason and Rubbish

appearance is of fundamental importance. The decisive issue is whether inherited differences with respect to mental and emotional make-up are associated with the ones we can detect.

It is admittedly clear that Jews have a religion and a tradition of their own. Since mediaeval times they have been compelled to live in towns. Often they could only settle in special quarters and could only carry on particular occupations. Since Jews were prohibited from performing many kinds of work, were not able to own land, could not enter the service of the State or become priests or soldiers, their only alternative has been to become business men, lawyers, and doctors. Because of anti-semitism the teaching profession is not attractive. A Jewish teacher has to face pupils whose parents may despise Jews, and the experience can scarcely be an agreeable one; but if the Jew is a lawyer, a doctor, or a business man, he need not come into daily contact with people who adopt an attitude of hatred towards him. Because of this social situation, the Jews have got a peculiar occupational orientation, and therewith traditions of their own. A Jew knows that he must reckon with the possibility that some of the people he meets are anti-semites, and in civilised conditions he cannot always and immediately be clear about which of those with whom he does come into contact have more or less active antipathy for his kind. Inescapably, he treats Gentiles with some suspicion and bears himself with

Race, Reason and Rubbish

aloofness. Naturally, he can feel greater confidence for other Jews.

The outcome is that Jews cultivate a special attitude to Gentiles and develop traits peculiar to it. In such circumstances it is difficult to conclude that they must share inherited characteristics which distinguish them more or less decisively from Christians. Psychological characters which are hereditable resolve themselves into intelligence and temperament, when these terms are interpreted in the broadest sense. With reference to intelligence no clear-cut differences can be attributed to heredity. At best there are differences of degree. There is no distinctively Jewish, in contradistinction to Christian or Aryan way of thinking. We can think logically, illogically, or not at all, and the main possibilities are then exhausted. So the real question is whether the capacity for thinking logically is better or worse among Jews than among Gentiles.

There are certainly no grounds for thinking that the Jews are worse—rather the contrary. Among men of science there are comparatively many Jews. Of one hundred and eighty-three Nobel prize winners up to the year 1936, fifteen were Jews and seven half-Jews, i.e. about 12 %. Such a figure certainly represents more than twice the proportion of Jews in the communities involved, and therefore means that the likelihood that a Jew will become a Nobel prize man is more than twice as great as the likelihood that a Gentile will do so. This

Race, Reason and Rubbish

does not necessarily prove an inborn superiority of the Jews in connexion with scientific research. It can partly be explained by the fact that they apply themselves so much to intellectual occupations. Because they have been forbidden to be farmers or soldiers or clergymen, they have been squeezed into real university education. So some have been able to go further and become scientists. That they belong so largely to the prosperous classes of western nations works in the same direction. We cannot therefore find sufficient scientific basis for the conclusion that Jews are inordinately intelligent. On the other hand, there is absolutely no reason for believing the opposite, and it is usual for anti-semites themselves to assert it. The truth of the matter is that there is a wide range of variation from genius to mental defect among Jews and Gentiles alike. The majority of both are intellectually mediocre, and the average level is probably the same for each.

What people more often try to make out is that Jews have less desirable moral qualities. If we turn to the statistics of crime we get no established basis for this. Comparatively speaking, Jews are seldom prone to violence, but are more addicted to embezzlement, bankruptcy, and the like. If we pay regard to the fact that comparatively many of them are business men we cannot deduce any difference. Criminality is not more common among Jews than among Christians.

It is sometimes alleged that Jews are more ruthless

Race, Reason and Rubbish

and unprincipled than other people. Certain ground for saying so does not exist. There is no single investigation which could show such a difference. Among Jews, as among the rest of us, there is a wide range of variation from saints at one extreme to criminal irresponsibility at the other, with a middle region in which different grades of selfishness and altruism are blended. Traits of this sort are naturally difficult to assess, but while we have no means of measuring them we ought to refrain from asserting the existence of dissimilarities, when we have no evidence based on research, and no proof of an average difference.

On the whole the moral standard depends on the intelligence of the parties concerned. A person who is excessively tight-fisted, egotistic, and ruthless, is liable to be avoided and gets paid back for it. So an intelligent person does not behave in this way. In the long run it is dangerous for a business man to be dishonourable. If he wants his affairs to prosper he must behave as honourably as others, because no one likes to play ball with a person who cheats. People used to urge that Jews must be morally deficient because they are now avoided as a group. The truth is otherwise. As individuals they are enterprising and have reached positions of prosperity to a comparatively remarkable extent. Their success points to the conclusion that they behave dependably and are not inordinately unscrupulous.

There is no real scientific basis for the statement that

Race, Reason and Rubbish

Jews are inferior, and unprejudiced discussion does not lead us to the conclusion that this opinion is a plausible one. We therefore ask ourselves how hatred of Jews can be explained. The explanation is really much the same as for the periodical flare up of hate between nations. All nations with common boundaries have been at war and have hated one another at some time. People do not start wars unless they have the prospect of victory. When they are equally strong they reason and compromise about conflicting interests; but if one is weaker it becomes the object of attack and therefore of hatred. The Jews have always been a minority. So it has always been possible to hate them.

National hatred is partly based on a gregariousness which holds neighbours together, but also carries with it antagonism to others, and above all to foreigners. A class of school children stick with one another against new pupils and there is often antipathy between schools. At a more civilised level we shall recognise that such antipathies are childish and serve no human use. So far as Jews are concerned hostility gives an advantage to their rivals. If they can get rid of Jewish competition, they themselves have a better chance of getting on. If they can make away with Jewish doctors those who are left can share a larger clientele, and if Jewish business men are forbidden to sell, others can sell more. So anti-semitism has always had an appeal for the egotism of the Gentiles.

Race, Reason and Rubbish

A tradition such as anti-semitism has extraordinary vitality. Below the Pyrenees there is a small group of people, the *cagots*, who are despised and persecuted. They can carry on only work which is ill-esteemed, for instance as lumberjacks, and they can marry only among themselves. This is not a matter of legal enactments but a social boycott of very early date. The origin of the boycott has been traced back, and the explanation turns out to be that the cagots are descended from lepers. This was several centuries ago and people have forgotten why the group was set apart, but the tradition is as much alive as ever. Every child in the neighbourhood has planted in his brain the notion that cagots are a bad lot, and every child when grown up will treat them as such and teach his or her children to do so. Social agencies of this sort obviously play an essential role in anti-semitism.

We should remember that Jews have learned to stick together in the same hard school; and they are often upbraided for doing so. If they did not stick together we should presumably detect evidence of moral baseness which would serve the uses of Jew-baiting just as well. Under the pressure of persecution and anti-semitism a Jewish nationalism has blossomed into confused notions about the peculiar genius of the Jewish people. Not unnaturally, Jews themselves believe that this national or racial genius expresses itself in highly commendable attributes. There is as little to be said for this opinion as for Jewish inferiority.

Race, Reason and Rubbish

Even if anti-semitism were mainly founded on correct views, and even if the average Jew were more dishonourable or more unscrupulous than other people, there would still be no reason for boycotting or persecuting every Jew. As persons we all prefer to be judged according to our own deserts and shortcomings. We regard it as outrageous if we have to suffer for the sins of our fathers or poor relations. We do not consider the skeleton in the cupboard a good reason for considering every member of the family as worthless. Why should we not treat the Jews in the same way as we ourselves would wish to be treated? Why should we persecute Jews without regard to whether they are clever and honourable or stupid and dishonest? Justice demands that each should be treated as he or she deserves, and that the Jew as an individual should not be judged by his ancestry without regard to his personal qualities.

To be sure, the Jewish question is a complex one, and a fundamental treatment of all the problems which arise in this connexion would call for a separate volume. Here we have only been able to discuss it briefly from the standpoint of heredity. It is enough to say that there is no scientifically acceptable proof that Jews on the average are worse than other folk. So far as inherited traits are concerned, there is absolutely no reason for maintaining that Jews represent a special type. Arguments brought forward from anti-semitic sources are without scientific justification.

CONCLUSION

In what has gone before we have talked about different processes which can influence the hereditary make-up of a population. When mating is haphazard, and the fertility of people of dissimilar types is approximately the same, no changes of this kind occur from one generation to another. To some extent it is really true to say that human marriages are haphazard; but we have also seen that there are five processes which bring about changes of the inherited characteristics of a people in the course of generations.

Only two of the five lead to essential alterations of the inherited stock-in-trade of the community; or, to fall back on a well-worn metaphor, only two involve a change of national income. *Mutation* brings about the appearance of new genes, but this phenomenon occurs so seldom that mutations have no significance on a short view of history, and a greater or smaller frequency of mutations does not mean that the hereditary make-up of a population mating at random will change much. Through the other process called *selection*, i.e. differences in the effective rate of reproduction, striking changes involving the occurrence of inherited traits can be brought about. If it involves relatively common characters the effect can be rapid and powerful, and even small differences of fertility can then be significant. For very rare recessive characters or characters inherited in a complicated way the effect of selection is insignificant.

The three remaining processes lead to no change in the stock-in-trade of genes. They involve nothing more than how the same gross national income is distributed among different sections of the community. The genes are dealt out in a new way. All these processes work in the same direction,[1] to produce an increase in the proportion of homozygotes and a decrease in the proportion of heterozygotes. *Inbreeding*, which is one of them, has little effect on the whole. When it involves common characteristics its influence is utterly negligible. When it involves very rare ones its effect is greater, but when this is so, the characters themselves are relatively unimportant. With regard to a second process, *assortative mating*, we lack materials for a very definite pronouncement. The personal opinion of the author is that any influence it may have had in earlier times had become exhausted until recently, but that the choice of like for like may now have begun to play a more important part in the distribution of very rare characters. On the other hand, it may well be that the significance of assortative mating is greater than we suppose. The last of the processes with which we have to reckon is the *isolate effect*. We are now going through a period when the frontiers of isolates are breaking down. This must

---

[1] In so far as assortative mating covers the possibility of preference for *opposites*, this is not true (*vide* p. 175).

Race, Reason and Rubbish

produce, and must have produced, a reduction in the proportion of rare recessive traits. With regard to more common characteristics such as stature or intelligence its effect may be very great, but we must get more material to justify a dogmatic statement about this.

In such circumstances the two fundamental questions of human genetics are *first*, the significance of selection arising from different fertility in social groups; and *second*, the dissolution of isolates because of migration to the towns, industrialisation, and the growth of modern communications. With regard to both, we are faced with problems which are by no means fully solved. We have no basis for very decided views and cannot therefore suggest practical applications based on scientific conclusion so far reached. Intensive research is necessary before we reach the point when there is sufficient reason for paying much attention to the political consequences of race biology from a constructive point of view.

The issues which now extend to the plane of politics are in part concerned with proposals for sterilising defective individuals and in part with opinions directed against Jews, Negroes, or other subject races. Neither the one nor the other has much significance from a strictly biological point of view. Sterilisation of defectives can never bring about results of great practical value. Meanwhile ill-considered handling of the problem may cause unnecessary suffering for a small section of individuals, but that is all it will do. If we sterilise a few we

Race, Reason and Rubbish

may avoid having to feed a trifling number of defective people. If we go too far, and particularly if we adopt compulsory sterilisation, we shall make an unnecessarily large number of people unhappy.

Practicable laws will not involve large numbers. Since sterilisation measures were promoted in Sweden the number of individuals sterilised in a single area has never exceeded four hundred. Curiously enough, nearly all of them have been women. Though the number of men below par must be about the same as the number of women, only about ten men in each year have been subjected to the operation. The truth is that men make up more than 50 % of mental defectives, a circumstance for which there is no explanation at present. During the first year after the new sterilisation laws were introduced in Germany fifty-six thousand persons were treated. This report was published in 1937. No later figures have been given out. There is no doubt that sterilisation is used much more in Germany than in Sweden or America, but since later records have not been published in the intervening period, we lack definite bases for a further verdict on the situation.

From a coldly rational point of view the position with reference to anti-semitism is in one way on all fours with the sterilisation issue. The groups concerned are relatively small, and this is specially so in some countries where Jews are subjected to harsh treatment. In another way the issues differ. We may suppose that

Race, Reason and Rubbish

such measures lead to a very great reduction of Jewish fertility and that anti-semitism influences Jewish minorities in other countries to some extent in the same way. The feeling of insecurity which being a Jew must entail, will presumably affect the production of children. In earlier times Jew-baiting and pogroms could not have had so much effect, because opportunities of family limitation were not available as they now are. So we may forecast that anti-semitism will gradually die for lack of material, even if it may take several generations.

We used to talk about positive and negative eugenics. Positive eugenics meant patronising reproduction among some social groups, and negative eugenics meant rationing it among others. In principle, the two issues are identical, because we cannot diminish reproduction in some groups without increasing it proportionally in others, and *vice versa*. In practice negative eugenics was mainly a matter of sterilisation and similar measures discussed in the foregoing. We have now seen that the anticipations raised by them are exaggerated. In one way the line of argument is right enough, because we do not want people who are physically, morally, or mentally undersized to go on breeding, but so far we have been content with inaccurate calculations of the effects which it is possible to get. The familiar arguments are right in outline; but exact analysis has shown that the measures proposed have no practical importance.

In principle positive eugenics implies that it is a good

Race, Reason and Rubbish

thing to stimulate fertility among specially talented people, but we have not found the mouse to bell the cat. This is specially due to the difficulty of reaching unanimity about what we mean by talent. One of us thinks that capacity for thinking logically is the most important thing in the world. Another considers that artistic gifts are the most significant ones. A third believes that what is called practical competence and skill, or a fourth that loyalty to a particular creed is the only way to salvation. With regard to deficient individuals who suffer from disease or congenital defect agreement is possible, because no one holds that it is a good thing to get more morons in the next generation than we now have. On the other hand, we cannot arrive at unanimity about what is meant by talent without a clear definition of individuals we regard as desirable. Perhaps it is best so. If those in power could decide what individuals there ought to be in future generations, we cannot be certain that the result would be felicitous.

Besides, it ought to be clear that we do not merely need individuals of one type. It takes all sorts to make a world. For the good government of a community we need persons with the irritating propensity of thinking independently, criticising current conditions, and demanding thorough-going changes of social organisation. We need persons with artistic gifts. We also need people with practical competence, even if they may not always have a very strong aptitude for logic or scientific dis-

Race, Reason and Rubbish

crimination. In addition, we must have less gifted people who are not wanted for their intellectual equipment but are still needed for the necessary, simple, monotonous tasks of community life. In the long run we may now hope to reach better social organisation and a technical level at which machines lift most of the dull monotonous toil from the shoulders of mankind, but we have not yet reached this state of affairs. Meanwhile it is not possible to define the proportions in which we need different types of people. So presumably we have grounds for being glad that it is not possible to do more about deciding the hereditary make-up of posterity.

Possibly we shall come to this. The growth of science is swift and cannot be restrained. To a large extent our future depends on how far and how swiftly scientific discoveries are made, and in what fields they turn up. In some measure we can stimulate its growth collectively, because scientific work depends on the means at its disposal. As we realise its potentialities more and more, its evolution will proceed faster, and we can naturally guide the process by encouraging particular branches, above all those from which we can expect socially useful results.

Meanwhile it is a fact about scientific work, that it is extraordinarily difficult, not to say impossible, to be certain what results we shall get from a particular investigation or work in a particular field. Researches which seem to have an exclusively theoretical interest, suddenly lead to unexpected consequences of practical importance. Within the world of science we have therefore been accustomed to maintain that the result should be valued without regard to its immediate practical worth, and that we should seek the truth for its own sake. So phrased, this is perhaps an exaggeration. None the less we must strive for truth alone in scientific investigation. Any attempt to control science and suppress its results will be luckless in the long run. Because many opinions which are politically alive and ostensibly connected with race biology are also wrong, nothing is more certain that they will disappear eventually. When scientific research has got further by well planned and well conducted work, there will be no opportunities for the majority of such opinions to remain fashionable. In the last resort no opinion can stand up to the unanimous verdict of science.

Bei Fragen zur Produktsicherheit wenden Sie sich bitte an:
If you have any questions regarding product safety,
please contact:

Walter de Gruyter GmbH
Genthiner Straße 13
10785 Berlin
productsafety@degruyterbrill.com